Mathematics for Electricians and Electronics Technicians

by Rex Miller and Martin R. Miller

THEODORE AUDEL & CO.
a division of
G. K. HALL & CO.
Boston

FIRST EDITION
85 86 87 88 / 4 3 2 1

For information, address G. K. Hall & Co., 70 Lincoln Street,
Boston, MA 02111

Manufactured in the United States of America

Miller, Rex, 1929-
Mathematics for electricians and electronics technicians.

Includes index
1. Electric engineering—Mathematics.
2. Electronics—Mathematics. I. Miller, Martin R. (Martin Rex), 1958- II. Title.
TK153.M49 1985 621.3'01'51 85-7495
ISBN 0-8161-1700-4

Contents

CHAPTER 6

CHAPTER 7

CHAPTER 8

CHAPTER 9

CHAPTER 10

Preface

The goal of this book is to provide an understanding of the mathematical concepts involved in the everyday life of an electrician or electronics technician. The simple math problem eludes some, but with a little time and effort it is possible to obtain a working knowledge with this simple 1-2-3 approach.

All the problems can be solved with a calculator. Nothing more complicated than a few trig problems are encountered, and all you have to do is hit the keys on the calculator to obtain the correct answer.

The applied principles of electricity are needed by everyone who will be working with communications equipment, electrical wiring, or any of a number of the industry's latest devices. Without this firm foundation, you are working with holes in your background. This book attempts to fill in the holes and provide a background that is useful for anyone who may want to learn more about his job or learn just for the fun of it.

Primarily, then, the purpose of this book is to provide the needed information for an electrician or electronics technician. It will also come in handy for beginning engineers, shop foremen, industrial arts teachers, electrical and electronics equipment salespeople, computer technicians, maintenance persons, and designers.

CHAPTER 1

Some Basic Math Concepts

Arithmetic is the art of dealing with real numbers. It deals with the addition, subtraction, multiplication, and division of numbers. If you want to extend the manipulation of numbers and observe the other properties that numbers possess, it is necessary to take a look at a higher step on the ladder of mathematics. Algebra is that next step.

Algebra is the branch of mathematics that treats of the relations and properties of numbers by means of letters, signs of operation, and other symbols, including solutions of equations, polynomials, continued fractions, and other mathematical phenomena. Algebra plays a fundamental role in the solution of electrical problems. It can be used to express abstract concepts in simple terms. These abstract concepts are important in working with electricity and electronics.

ELECTRONICS AND MATHEMATICS

Electricity is used in making electronics functional. Without electricity, there is no electronics. Therefore, some basic knowledge of electricity is required for a better understanding of electronics.

Electronics deals with symbols; so does electricity. The symbols for voltage, resistance, and current are used in formulas to express a relationship between or among the values of current and resistance, voltage and resistance, current and voltage, and other properties of a circuit. In order to do this in a formula, it is necessary to identify the symbols and then place them in a relative position.

The positioning of the symbols for voltage, current, and resistance represents the interaction of these three properties of electricity. Algebra can be used to express the relationship easily. It can also be used to determine a missing value if two of the values in a circuit are known. This is but one of the many aspects of algebra being used in extending the knowledge of electricity in circuits.

SYMBOLS

The best place to start with algebra and electricity is Ohm's law. Ohm stated that the electric current is inversely proportional to the resistance in any circuit. This means that if we want to write it in a formula, we have to express each quantity in an abbreviated form. This we do by saying that voltage is E, current is I, and resistance is represented by R. *If the current in a circuit is equal to the voltage divided by the resistance*, then we can state Ohm's law in a formula:

$$I = \frac{E}{R}$$

Algebraic Signs

Algebra uses the same signs as simple arithmetic: $-$ for subtraction; $+$ for addition; $\times$ for multiplication; and $\div$ for division. However, the multiplication sign is usually omitted and the two letters are brought together such as in Ohm's law, where $E = I \times R$ becomes $E = IR$. Sometimes, the $\times$ is replaced with a $\cdot$ symbol. Thus, $I \cdot R$ means the same thing as IR or $I \times R$.

More complicated formulas also make use of the algebra shortening of the multiplication sign. For instance, when the formula for inductive reactance is used, it leaves out this sign: $X_L = 2\pi FL$. The 2 times π times F times L is understood without the $\cdot$ or the $\times$.

Mathematical Order of Sign Usage

Multiplication, division, addition, and subtraction are to be performed in a specific order when they are used in an equation or problem in which two or more of the mathematical signs are used.

1. Do all the multiplications, in order, from left to right.

2. Do all divisions, in order, from left to right.
3. Do all additions and subtractions, in order, from left to right.

For example:

$$5 \times 3 + 8 - 6 + 20 \div 4 = ?$$

1. Multiply $5 \times 3 = 15$
2. Divide 20 by $4 = 5$
3. Subtract 6 from $8 = 2$
4. Then the 15 is added to 5 to produce 20.
5. This 20 is added to the 2 to produce 22, the final answer.

METHODS OF EXPRESSING ALGEBRAIC RELATIONSHIPS

Two methods are used to express algebraic relationships. The *numerical* expression uses signs and numbers only. For instance, $10 - 6(3 + 2)$ is such an expression. However, if you express relationships by means of letters and numbers or letters alone, it is called a *literal* expression. An example would be I^2Z or $\frac{E^2}{R}$.

Obtaining the Product

The *product* is the result of multiplying two numbers. For instance, 5×10 is 50. Here 50 is called the product. This is the same as in simple arithmetic.

Obtaining the Factor

The *factor* is obtained when two or more numbers are multiplied together. Any combination of the numbers multiplied together produces a factor. For instance, if you have a result of $4ab$, then the factors of $4ab$ are 4, a, and b. The factors can also be $4a$, $4b$, a, b, and ab.

Identifying Coefficients

Coefficients are when given terms represent the product of two or more factors. That means that each factor is the coefficient of the other or others. For example, in the term $4a^2b^2c^3$, 4 is the coefficient of $a^2b^2c^3$, a^2 is the coefficient of $4b^2c^3$, b^2 is the coefficient of $4a^2c^3$, and c^3 is the coefficient of $4a^2b^2$.

There are two names for coefficients. The number (here 4) is called a *numerical coefficient*, but we often just refer to it as the coefficient.

Subs, Supers, and Primes

We use subscripts to designate a particular resistor, capacitor, inductor, or transformer. R_1, C_1, L_1 are all examples of the *sub*script used to designate a particular device. It comes in handy in the shorthand method of representing devices located on a schematic. It is also very useful in formulas. They are read as "R sub one" or "C sub one."

We use *super*scripts in another way. The term x^2 means "x squared." The 2 is a super and goes above the letter. The term x^3 means "x cubed." These superscripts are also known as *exponents*. They show up on the calculator also in the exponent window on the right-hand side of the calculator.

A prime (′) indicates a particular value. E' is read "voltage prime," I' is "current prime," and R' indicates "resistance prime". This is sometimes read (mistakenly) as E^1 or R^1. But this isn't too much of a problem since E^1 or R^1 or even I^1 has no meaning.

It is also possible to use the symbol ″ to indicate seconds. E'' would be read as "E second" or "the one following E'". So if you are using only two voltages, it is possible to designate one as E' and the other as E''; or they become the prime voltage and the second voltage.

Finding Numerical Value

In algebra numbers and letters are used in equations. It is possible to solve problems or equations thus represented by knowing the value of some of the letters. These values are then substituted into the equation and the answer is obtained by normal mathematical procedures. For example:

Find the value of $4xy$.
The value of x is 5.
The value of y is 4.

Substitute and you get the expression $4\times5\times4$. Multiply to obtain the answer, 80.

Find the value of $41x-6ab$.
The value of x is 2.
The value of a is 3.
The value of b is 4.

Substitute the values of the letters into the expression to obtain $41\times2-6\times3\times4$. $41\times2=82$ and $6\times3\times4=72$. The next step is to subtract the 72 from the 82, or $82-72=10$. Therefore the value of the expression is 10.

Find the value of $\frac{X_L}{X_C}-4R$

X_L is equal to 500.
X_C is equal to 10.
R is equal to 5.

To find the answer substitute:

$$\frac{500}{10}-(4\times5)=50-20=30$$

Exponents

As mentioned before, exponents are numbers placed above a number or letter and represent the number of times the value of the number or letter is to be multiplied by itself. A^2 is read "A squared." It means A is multiplied by itself twice, or A times A.

If the number 4 is raised to a power of 4, it would be written as 4^4 and would mean that 4 is multiplied by itself 4 times, or $4\times4\times4\times4=256$.

The following are examples of exponents: a^2, x^3, z^6, 10^2, 15^3, and a^2b^3. Most of today's calculators will have "EE" or "E EX" to indicate the exponent key. "x^y" is used by calculators to allow you to enter the number and then the power to which it is to be raised. The "y" indicates the exponent.

Using the Square Root or Radical Sign

The square root sign ($\sqrt{\ }$) has the same meaning in algebra as it does in arithmetic. The radical sign, however, can be used to indicate the root of any number. For instance, $\sqrt[3]{a}$ represents the cube root of a, and written as $\sqrt[4]{5}$, it represents the fourth root of 5. The small number in the radical sign is called the index of the root. Any number can be placed in the radical sign and the index number can be assigned accordingly as needed.

Like and Unlike Terms

There are like or similar terms used in algebra to indicate expressions where the same or similar terms are used. Examples are $4a$, $5a$, $6a$. All have the a as a common term or like term. The terms $6a^2bz$, $5a^2bz$, and $-a^2bz$ are all similar since they have the a^2bz in common.

Unlike terms are also used. They may also be called dissimilar terms. Both expressions mean the same. Examples are $6abz$, $5a^2bx$, and $6y$. They share no common or similar term; hence the phrase *dissimilar terms*.

A term with a $-$ preceding it is referred to as a *negative term.* A term with a $+$ preceding it is a *positive term.* Keep in mind that the word *term* applies to a grouping of numbers or letters or a combination of letters and numbers.

Other Terms

As we progress into the use of mathematics in electricity and electronics, you will be able to grasp the meaning of the equations and types of procedures to use. Each step will be explained to you as needed. At this time there is no need to become involved in the theoretical math. We are primarily interested in the electrical application of mathematics. That is why you will be given examples and explanations as you proceed to each topic.

These topics have been designed and placed in a pattern or sequence to allow you to meet the basic knowledge requirements of the electrician or the electronics technician. If a firm foundation is laid to begin with, it is possible to understand much more of the special terms and their meaning as they are encountered in daily practice.

CHAPTER 2

Resistors and Resistance

WIRE SIZE AND RESISTANCE

In 1857 the American Wire Gage (AWG) was developed by J. R. Brown as a method for measuring wire size. It was also known for a time as the Brown and Sharpe (B&S) gage, since that was the name of the company that Brown owned. The basis for the gage is found in a simple mathematical law. The gage is formed by specifying the diameters of two of the wire sizes with the intermediate diameter of the gage numbers between the two specified numbers formed by a geometric progression. There are 40 gage numbers plus 4 zero designations. There is 0000, or 4-ought; 000, or three-ought; 00, double-ought; and 0, or ought. The 0000 is the largest size for the wire gage and #40 is the smallest, so you have to keep in mind that *the larger the number, the smaller the diameter of the wire.* The larger numbers are written as 4/0, 2/0, etc.

The 4/0 diameter is 0.4600 in. Number 36 has a 0.0050-in. diameter. There are 38 sizes between these two gage numbers. The ratio of any diameter to the diameter of the next larger number is the 39th root of the ratio of the two specified diameters, or 1.1229322.

Table 2-1 shows wire gage numbers between 4/0 and 40. There are a couple of things you should notice in order to be aware of the wire size and gage number relationships.

1. An increase of three gage numbers doubles the area and divides the resistance by two.
2. An increase of six gage numbers doubles the diameter and divides the resistance by four.

Keep in mind that the diameter in *mils* is converted to inches by moving the decimal point three places to the left. Or, in other words, the 5 mils shown for #36 wire equals 0.005 in.

The second column in Table 2-1 gives the diameter in mils. The third column gives the *circular mils*. Circular mils are the mils squared. For example, the diameter of 4/0 wire is 460 mils and the circular mils are 460 × 460, or 212,000 circular mils when rounded off.

The fourth column gives the cross-sectional area of the copper wire in square inches. Note the fifth and sixth columns, where the ohms per 1,000 feet are given for both 25°C (77°F) and 65°C (149°F), respectively. This can be used to calculate the relative resistances of wire being used inside a building in normal temperatures and inside a transformer or other device where temperatures may reach as high as 65°C (149°F).

In some instances it will be necessary to find the resistance of the copper wire for long distances. In telephone lines and power transmission lines the length of the wire is in miles rather than feet. The seventh column gives the ohms per mile. The last column provides the weight of the copper wire. Shipping and handling wire becomes a problem when it is miles in length. For instance, 4/0 wire weighs 641 pounds per 1,000 feet. This can be rather bulky and hard to handle, so in most instances, it will not be shipped in 1,000-foot rolls. At the other end of the column you find that #40 wire weighs only half an ounce per 1,000 feet.

Take a look at Table 2-2. It gives the current capacity at 700 CM/ampere, or 700 circular mils per ampere. If you are going to wind transformers or inductors, you need to know the turns per inch and turns per square inch. Formvar is an insulation that keeps the wire from shorting out when it is wound one layer on top of the other.

USING THE TABLES

As mentioned before, tables have a number of uses. For instance, tables can be helpful in putting information into a practical form and can aid in solving problems.

Table 2-1. Standard Annealed Solid Copper Wire
(American wire gage—B & S)

Gage Number	Diameter (Mils)	Cross Section		Ohms Per 1,000 Ft.		Ohms Per Mile 25°C (=77°F)	Pounds Per 1,000 Ft.
		Circular Mils	Square Inches	25°C (=77°F)	65°C (=149°F)		
0000	460.0	212,000.0	0.166	0.0500	0.0577	0.264	641.0
000	410.0	168,000.0	0.132	0.0630	0.0727	.333	508.0
00	365.0	133,000.0	0.105	0.0795	0.0917	.420	403.0
0	325.0	106,000.0	0.0829	0.100	0.116	.528	319.0
1	289.0	83,700.0	0.0657	0.126	0.146	.665	253.0
2	258.0	66,400.0	0.0521	0.159	0.184	.839	201.0
3	229.0	52,600.0	0.0413	0.201	0.232	1.061	159.0
4	204.0	41,700.0	0.0328	0.253	0.292	1.335	126.0
5	182.0	33,100.0	0.0260	0.319	0.369	1.685	100.0
6	162.0	26,300.0	0.0206	0.403	0.465	2.13	79.5
7	144.0	20,800.0	0.0164	0.508	0.586	2.68	63.0
8	128.0	16,500.0	0.0130	0.641	0.739	3.38	50.0
9	114.0	13,100.0	0.0103	0.808	0.932	4.27	39.6
10	102.0	10,400.0	0.00815	1.02	1.18	5.38	31.4
11	91.0	8,230.0	0.00647	1.28	1.48	6.75	24.9
12	81.0	6,530.0	0.00513	1.62	1.87	8.55	19.8
13	72.0	5,180.0	0.00407	2.04	2.36	10.77	15.7
14	64.0	4,110.0	0.00323	2.58	2.97	13.62	12.4
15	57.0	3,260.0	0.00256	3.25	3.75	17.16	9.86
16	51.0	2,580.0	0.00203	4.09	4.73	21.6	7.82
17	45.0	2,050.0	0.00161	5.16	5.96	27.2	6.20
18	40.0	1,620.0	0.00128	6.51	7.51	34.4	4.92

Table 2-1. Standard Annealed Solid Copper Wire (Cont'd)
(American wire gage—B & S)

Gage Number	Diameter (Mils)	Cross Section		Ohms Per 1,000 Ft.		Ohms Per Mile 25°C (=77°F)	Pounds Per 1,000 Ft.
		Circular Mils	Square Inches	25°C (=77°F)	65°C (=149°F)		
19	36.0	1,290.0	.00101	8.21	9.48	43.3	3.90
20	32.0	1,020.0	.000802	10.4	11.9	54.9	3.09
21	28.5	810.0	.000636	13.1	15.1	69.1	2.45
22	25.3	642.0	.000505	16.5	19.0	87.1	1.94
23	22.6	509.0	.000400	20.8	24.0	109.8	1.54
24	20.1	404.0	.000317	26.2	30.2	138.3	1.22
25	17.9	320.0	.000252	33.0	38.1	174.1	0.970
26	15.9	254.0	.000200	41.6	48.0	220.0	0.769
27	14.2	202.0	.000158	52.5	60.6	277.0	0.610
28	12.6	160.0	.000126	66.2	76.4	350.0	0.484
29	11.3	127.0	.0000995	83.4	96.3	440.0	0.384
30	10.0	101.0	.0000789	105.0	121.0	554.0	0.304
31	8.9	79.7	.0000626	133.0	153.0	702.0	0.241
32	8.0	63.2	.0000496	167.0	193.0	882.0	0.191
33	7.1	50.1	.0000394	211.0	243.0	1,114.0	0.152
34	6.3	39.8	.0000312	266.0	307.0	1,404.0	0.120
35	5.6	31.5	.0000248	335.0	387.0	1,769.0	0.0954
36	5.0	25.0	.0000196	423.0	488.0	2,230.0	0.0757
37	4.5	19.8	.0000156	533.0	616.0	2,810.0	0.0600
38	4.0	15.7	.0000123	673.0	776.0	3,550.0	0.0476
39	3.5	12.5	.0000098	848.0	979.0	4,480.0	0.0377
40	3.1	9.9	.0000078	1,070.0	1,230.0	5,650.0	0.0299

Table 2-2.

AWG (B&S) Gage Number	Current Capacity @ 700 CM/Amp.	Single Formvar			Heavy Formvar		
		Diam.	Turns/In.	Turns/In.^{2}	Diam.	Turns/In.	Turns/In.^{2}
No. 8	23.5859	0.1306	7	49	0.1323	7	49
No. 9	18.7039	0.1165	8	64	0.1181	8	64
No. 10	14.8329	0.1039	9	81	0.1055	9	81
No. 11	11.7630	0.0927	10	100	0.0942	10	100
No. 12	9.3285	0.0827	12	144	0.0842	11	121
No. 13	7.3978	0.0738	13	169	0.0753	13	169
No. 14	5.8667	0.0659	15	225	0.0673	14	196
No. 15	4.6525	0.0588	17	289	0.0602	16	256
No. 16	3.6896	0.0524	19	361	0.0538	18	324
No. 17	2.9260	0.0469	21	441	0.0482	20	400
No. 18	2.3204	0.0418	23	529	0.0431	23	529
No. 19	1.8402	0.0374	26	676	0.0386	25	625
No. 20	1.4593	0.0334	29	841	0.0346	28	784
No. 21	1.1573	0.0299	33	1089	0.0310	32	1024
No. 22	0.9178	0.0266	37	1369	0.0277	36	1296
No. 23	0.7278	0.0238	42	1764	0.0249	40	1600
No. 24	0.5772	0.0213	46	2116	0.0223	44	1936
No. 25	0.4577	0.0190	52	2704	0.0200	50	2500
No. 26	0.3630	0.0169	59	3481	0.0179	55	3025
No. 27	0.2879	0.0152	65	4225	0.0161	62	3844
No. 28	0.2283	0.0135	74	5476	0.0145	68	4624
No. 29	0.1810	0.0122	81	6561	0.0131	76	5776
No. 30	0.1436	0.0108	92	8464	0.0116	85	7396
No. 31	0.1139	0.0097	103	10609	0.0104	96	9215

Example 1
What is the resistance of 5,000 feet of copper wire if the temperature is 77°F and the wire size is 4/0?

1. What do you want to find? *Resistance.*
2. What is given? *Temperature, length,* and *wire size.*
3. Check Table 2-1 for wire size first. It's in the first column. Move over to the temperature you want under the heading "Ohms Per 1,000 Ft." It reads 0.0500 for 1,000 feet. But you want the resistance for 5,000 feet.
4. So, multiply the 0.0500 by 5: $0.0500 \times 5 = 0.250$ ohms resistance for the 5,000 feet of copper wire at 77°F.

Example 2
What is the weight of 5,000 feet of copper wire that measures #2 on the AWG gage?

1. Check out the tables. Find the right one. In this case it is Table 2-1.
2. Find the gage number in the first column.
3. Follow this out to the "Pounds Per 1,000 Ft" column, the last column.
4. Take the 201 pounds and multiply it by 5 since you have 5,000 feet of wire.
5. $201 \times 5 = 1{,}005$ pounds for the 5,000 feet.

Example 3
What is the current-carrying capacity of #14 copper wire?

1. Look for the correct table with this information. In this case it is Table 2-2.
2. Find the gage number. Run your finger across the table until you find the column for current capacity.
3. The current capacity of #14 copper wire is 5.8667 amps.

Example 4
What is the difference in diameter between single and heavy Formvar insulation on #10 copper wire?

1. Table 2-2 is the one with this information.
2. Read the column for single Formvar and record it. In this case it is 0.1039 in.
3. Read the column for heavy Formvar and record it. In this instance it is 0.1055 in.
4. To find the difference, subtract the single-Formvar diameter from the heavy-Formvar diameter, or $0.1055 - 0.1039 = 0.0016$ in. difference.

This 0.0016 doesn't mean much as a single number, but it makes a difference when you are winding hundreds of feet over a bobbin to be used as a transformer or electric motor. The space it takes may make a difference between whether or not the finished winding will fit into its intended space.

PROBLEMS

1. What is the diameter (in inches) of #30 copper wire?
2. What is the diameter (in mils) of #30 copper wire?
3. What is the cross-sectional area of #32 wire (in circular mils)?
4. What is the cross-sectional area of #40 copper wire (in square inches)?
5. What is the ohms/1,000 feet for #16 copper wire?
6. What is the resistance of 50,000 feet of #40 copper wire?
7. What is the resistance of 5 miles of #20 copper wire at 25°C?
8. How much does 10,000 feet of #14 wire weigh?
9. How much does 500 feet of 4/0 wire weigh?
10. What is the current-carrying capacity of #20 wire?
11. How many *turns/in.* can you get from #16 copper wire coated with heavy Formvar?
12. How many *turns/in.*^{2} can you get from #28 copper wire coated with single Formvar insulation?

RESISTANCE AND RESISTORS

Resistance serves three roles in a circuit. These roles have direct relationships to the proper function of a circuit.

1. Resistance converts electrical energy into heat and is thus used in electric heaters, stoves, and toasters.
2. Resistance can be used to limit the flow of electrons within a circuit. This makes it possible to ensure that the proper voltage or current reaches a certain point in an electrical circuit.
3. Resistance provides a load within circuits. A load is any device that uses electric current or power. Loads may be used to control current or voltage within a circuit.

When there are special reasons to add resistance to a circuit, resistors are used. A *resistor* is a device that limits, or controls, the flow of electrons through a circuit. A resistor adds resistance to a circuit. The amount of resistance needed in any circuit depends on the natural resistance already within the conductor itself. All conductors present some resistance to current flow. This resistance is determined by:

1. Length of the conductor.
2. Size of the conductor, including the diameter.
3. Temperature of the conductor.
4. Type of material used as a conductor.

Coefficient of Resistance

All materials have some resistance, and each material has its own resistance. Some of the materials used for their resistance properties are listed in Table 2-3. Note that the unit of measurement is the *ohms per circular mil.* The circular mil is 0.001 in. in cross-sectional area. *Mil* means "one-thousandth."

The circular *mil foot,* which is used here to show the coefficient of resistance, is simply the resistance of a piece of the material that is 1 foot long and measures 1 circular mil (0.001 in.) in cross-sectional area. The temperature is assumed to be 68°F (20°C).

Table 2-3. Coefficients of Resistance (in ohms per circular mil foot)

Aluminum	17.0
Carbon	4210.0
Constantan	295.0
Copper	10.4
Iron (pure)	60.2
Mercury	576.0
Nichrome	675.0
Platinum	59.5
Silver	9.9
Tungsten	33.1
Zinc	36.7

As you scan Table 2-3, note some of the following:

1. Copper has a low coefficient of resistance, but silver's is lower. Therefore, if you want a low line loss, you would use silver wire; but if you don't want the expense of silver, copper would be the next-best bet.
2. Note that aluminum does not have as small a coefficient of resistance as copper. It is less expensive than copper and has some other desirable characteristics that must be considered in some cases.
3. Carbon has the highest coefficient of the materials listed in Table 2-3. That means it can be used to make resistors or control-type devices.
4. Nichrome is a man-made alloy that can be used to produce heat. This material is used in toasters and ovens. It also has the ability to handle the higher temperatures generated by its own resistance to current flow.

Determining Resistance

The formula for computing the resistance of a length of material is

$$R = \frac{\rho \times L}{A}$$

R = resistance measured in ohms

ρ (Greek letter *rho*) = coefficient of resistance

L = length of the material in feet
A = cross-sectional area in circular mils

Problem Solving

Example 5

Find the resistance of 500 feet of 0.125 in. aluminum wire used in a factory at 68°F (20°C). Simply substitute the values in the problem into the formula given previously.

1. $R = ?$
2. $\rho = 17.0$ (taken from Table 2-3)
3. $L = 500$ feet
4. $A = 15{,}625$. Since the 0.125 in. equals 125 mils, the mils must be converted to circular mils area or d^2, the diameter squared. $125^2 = 15{,}625$.
5. $$R = \frac{17.0 \times 500}{15{,}625}$$
6. $R = 17 \times 500 = 8{,}500 \div 15{,}625$, which equals 0.544 ohms.

Example 6

Find the resistance of 10,000 feet of copper wire with a diameter of 0.0625 in. that operates in a temperature of 68°F (20°C).

Simply substitute the values in the problem into the formula. The individual parts of the formula must be identified first, as in Example 1.

1. $R = ?$
2. $\rho = 10.4$ (taken from Table 2-3)
3. $L = 10{,}000$ feet
4. $A = 0.0625$ in. diameter, which converts to 62.5 circular mils, Circular mils area is found by squaring the 62.5, equaling 3906.25.

5. $$R = \frac{10.4 \times 10{,}000}{3{,}906.25}$$

6. $R = 10.4 \times 10{,}000 = 104{,}000 \div 3{,}906.25$, which equals 26.624 ohms

There is one method that can be used to calculate how much *insulated* wire is in a spool or roll, and that is by using a very accurate ohmmeter, measuring the resistance, and then plugging the result back into the formula just used. If you know it is copper and you know the resistance as found by the ohmmeter, you can measure the wire diameter and then plug all the values into the formula and thus determine the length of the wire left on the spool or roll.

For example, you have a resistance of 10.295 ohms for a roll of aluminum wire and it is sitting outside, where the temperature is 68°F (20°C). What is the length of the wire on the coil if it measures 0.1285 in. in diameter?

1. The first thing you decide is the unknown. In this case it is the length of the wire, or L.
2. Then you have to decide which formula to use. That can be done by checking what is given.
3. You know that the temperature is 68°F, so you don't have to compensate for that.
4. You know the wire is aluminum, so its coefficient of resistance, obtained from Table 2-3, is 17.0.
5. You know the total resistance of the coil to be 10.295 ohms.
6. You also know the circular mils since the 0.1285 in. wire is 128.5 mils. The circular mils are found by 128.5^2, or 16,512.25.
7. Next, check the formula you used previously and you'll find all these values fit into it with the exception of the length, L. This can be found by the following method.
8. Substitute the values you know and solve
9. $$R = \frac{\rho \times L}{A}$$

10. $$10.295 = \frac{17 \times L}{16{,}512.25}$$

11. Multiply $17 \times L$ to get $17L$.
12. Multiply $10.295 \times 16{,}512.25$ to get 169,993.6138.
13. Next take the formula and arrange it so:

$$17L = 169{,}993.6138$$

14. Then L would equal 169,993.6138 divided by 17, which equals 9,999.624341 feet, or 10,000 feet for all practical purposes.

There are, of course, other ways to find the length of a coil of wire. One of them is to use the wire table available in most engineering source books.

Effect of Temperature on Resistance

The temperature is one of the variables which must be dealt with in figuring the resistance of any substance. In order to make the previous formula work properly, you must take into consideration the temperature factor. The temperature is *assumed* to be 68°F (20°C) if it is not otherwise noted. The tables are drawn up with the 68°F temperature as a reference.

Table 2-4 shows the temperature coefficients in degrees Fahr-

Table 2-4. Temperature Coefficients (in degrees Fahrenheit)

Aluminum	+0.00210
Carbon	−0.00014
Constantan	negligible
Copper	+0.00218
Iron	+0.00280
Mercury	+0.00050
Nichrome	+0.00010
Platinum	+0.00210
Silver	+0.00220
Tungsten	+0.00250
Zinc	+0.00210

enheit. Note in your scanning of the table that carbon has a negative coefficient.

The formula below takes into consideration the temperature variations that may exist where wire is used.

$$R = \frac{\rho \times L}{A}\,[1 + (T - 68)a]$$

R = resistance in ohms
ρ = coefficient of resistance
L = length in feet
A = cross-sectional area in circular mils
T = temperature in °F
a = temperature coefficient of the material

Problem Solving

Example 7

Find the resistance of 500 ft. of 0.1285-in. copper wire if the temperature is 98°F (37°C).

1. Determine what is known and what is unknown.
2. R is unknown.
3. $\rho = 10.4$ (taken from Table 2-3).
4. L is given at 500 feet.
5. The diameter of the wire will tell you the area in circular mils. The diameter is 0.1285 in. This means 128.5 mils. Circular mils are found by squaring the mils, or $128.5^2 = 16{,}512.25$.
6. Now that you know all the values for the first part of the formula, plug them into the formula and solve.
7. $$R = \frac{10.4 \times 500}{16{,}512.25}$$ $$R = 0.3149177125 \text{ ohms}$$
8. Next you have to compensate for the temperature difference. See what you know in this formula modification.
9. $T = 98°F$

10. $a = 0.00218$ (taken from Table 2-4)
11. So substitute to get $1 + (98 - 68)0.00218$.
12. Then subtract 68 from 98 to produce 30. This 30 is multiplied by the 0.00218 and added to the 1.
13. The answer is 1.0654, but you are not finished yet.
14. This 1.0654 must be multiplied by the resistance you obtained in step 7.
15. 0.3149177125 ohms × 1.0654 produces 0.3355133309 ohms.
16. So, you can see the 30° rise in temperature did make a difference in the total resistance of the wire. In the case of thousands of feet of wire it can make quite a difference.

Example 8

If you want to find the resistance of a length of wire without determining the temperature, you can look up the ohms per 1,000 feet or the current-carrying capacity or any number of different factors related to wire size and resistance. See Table 2-5. Suppose you have 1,500 feet of copper wire that measures 0.0159 in. in diameter and want to know what its resistance would be if you measured it. How do you find the resistance, which would also be one way of finding the length of an unknown piece of copper wire?

1. Look up the copper wire size in Table 2-5.
2. You find it is #26 wire and has a resistance of 40.8141 ohms per 1,000 feet.
3. You know that you have 1,500 feet. Therefore, simply multiply the resistance per 1,000 feet by 1.5 to get the answer.
4. $1.5 \times 40.8141 = 61.22115$ ohms
5. Of course, the accuracy is more than you want, so you can round it off to 61 or 61.2 ohms, depending on the ohmmeter you are using and its accuracy.

PROBLEMS

1. Find the resistance of 5,000 feet of 0.1285-in. diameter copper wire.

Table 2-5. Wire Table (Copper)

AWG (B&S) Gage Number	Diam. (in.)	Circular Mils	Ohms/ 1000′	Current Capacity @ 700 CM/Amp.	Single Formvar			Heavy Formvar		
					Diam.	Turns/In.	Turns/In.^{2}	Diam.	Turns/In.	Turns/In.^{2}
No. 8	0.1285	16509.65	0.6282	23.5859	0.1306	7	49	0.1323	7	49
No. 9	0.1144	13092.75	0.7921	18.7039	0.1165	8	64	0.1181	8	64
No. 10	0.1019	10383.02	0.9988	14.8329	0.1039	9	81	0.1055	9	81
No. 11	0.0907	8234.11	1.2595	11.7630	0.0927	10	100	0.0942	10	100
No. 12	0.0808	6529.95	1.5882	9.3285	0.0827	12	144	0.0842	11	121
No. 13	0.0720	5178.48	2.0027	7.3978	0.0738	13	169	0.0753	13	169
No. 14	0.0641	4106.72	2.5254	5.8667	0.0659	15	225	0.0673	14	196
No. 15	0.0571	3256.78	3.1844	4.6525	0.0588	17	289	0.0602	16	256
No. 16	0.0508	2582.74	4.0155	3.6896	0.0524	19	361	0.0538	18	324
No. 17	0.0453	2048.21	5.0634	2.9260	0.0469	21	441	0.0482	20	400
No. 18	0.0403	1624.30	6.3849	2.3204	0.0418	23	529	0.0431	23	529
No. 19	0.0359	1288.13	8.0512	1.8402	0.0374	26	676	0.0386	25	625
No. 20	0.0320	1021.53	10.1524	1.4593	0.0334	29	841	0.0346	28	784
No. 21	0.0285	810.11	12.8019	1.1573	0.0299	33	1089	0.0310	32	1024
No. 22	0.0253	642.45	16.1429	0.9178	0.0266	37	1369	0.0277	36	1296
No. 23	0.0226	509.49	20.3558	0.7278	0.0238	42	1764	0.0249	40	1600
No. 24	0.0201	404.04	25.6682	0.5772	0.0213	46	2116	0.0223	44	1936
No. 25	0.0179	320.42	32.3670	0.4577	0.0190	52	2704	0.0200	50	2500
No. 26	0.0159	254.10	40.8141	0.3630	0.0169	59	3481	0.0179	55	3025
No. 27	0.0142	201.51	51.4656	0.2879	0.0152	65	4225	0.0161	62	3844
No. 28	0.0126	159.81	64.8969	0.2283	0.0135	74	5476	0.0145	68	4624
No. 23	0.0113	126.73	81.8335	0.1810	0.0122	81	6561	0.0131	76	5776
No. 30	0.0100	100.50	103.1901	0.1436	0.0108	92	8464	0.0116	85	7396
No. 31	0.0089	79.70	130.1204	0.1139	0.0097	103	10609	0.0104	96	9215

2. What is the resistance of 25,000 feet of 0.0201-in. diameter copper wire?
3. What is the resistance of 500 feet of aluminum wire if the diameter is 0.250 in.?
4. What is the resistance of 500 feet of copper wire if the diameter is 0.250 inch?
5. Find the resistance of 25,000 feet of 0.1286-in. copper wire.
6. What is the resistance of 5,000 feet of 0.1144-in. aluminum wire?
7. What is the resistance of 5,000 feet of 0.0320-in. copper wire?
8. What is the resistance of 5,000 feet of 0.1144-in. copper wire?
9. What is the resistance of 5,000 feet of 0,0641-in. copper wire at 78°F (26°C)?
10. What is the resistance of 10,000 feet of aluminum wire if the diameter is 0.1019 in. and the temperature is 98°F (37°C)?

RESISTOR COLOR CODE

Wire-wound resistors are usually marked with numbers or letters. These markings indicate the power and resistance ratings for the devices.

Carbon-composition resistors are too small to be marked with numbers and letters. Instead, there are two different ways to tell their ohmic value. The power, or wattage rating, is indicated by physical size. The larger the carbon-composition resistor, the more power it can handle. As you come across carbon resistors in your work, you will have to learn to judge their wattage rating. This comes with experience.

The resistance of a carbon-composition resistor is marked with color bands. This color-band coding system is shown in Fig. 2-1. Take a moment to study the colors and their meanings.

These number values are applied to markings on all carbon-composition resistors. Slightly different systems are used to mark resistors rated at more than 10 ohms from those rated at less than

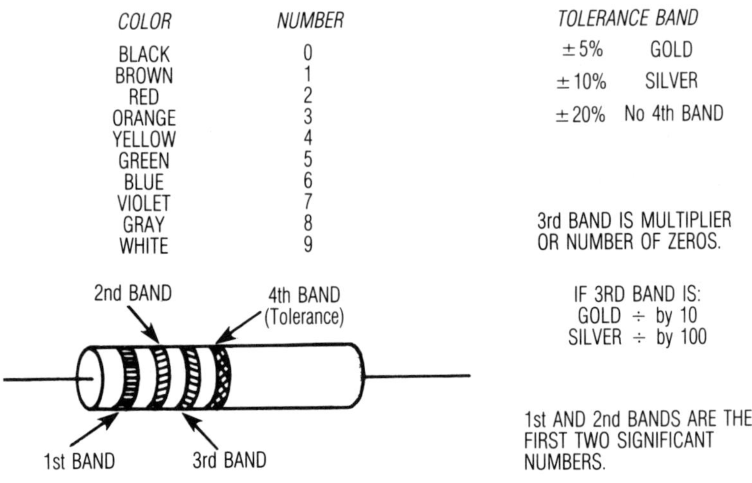

Fig. 2-1

10 ohms. You will learn first about markings of resistors rated at more than 10 ohms.

The color bands start at the left end of the resistor. You can identify the starting end because the color bands are closer to the end of the resistor. Read the number values from the starting end. The first band is the first digit of the number for the resistance rating. The second band is the second digit. The third band is a multiplier. The fourth band indicates the tolerance of the resistor and is either silver or gold in color.

For practice, look at Fig. 2-2. The red is valued at 2 (taken from the Fig. 2-1 listing). The second ring is yellow. This is 4. The next ring is orange and translates to 3 in the code. Since this is the multiplier, it means there are three zeros added after the 24. This tells you the resistor is 24,000 ohms in value. The fourth band indicates the tolerance. If it is silver, as indicated here, it means the 24,000-ohm resistor may read 2400 ohms (10%) *above* 24,000 ohms. It also means the resistor may read 2400 ohms (10%) *below* the 24,000 ohms and still be within tolerance. That means the resistor may be between 26,400 and 21,600 ohms and still be called a 24,000-ohm resistor. Tolerance is usually written as ±. This means "either + or −."

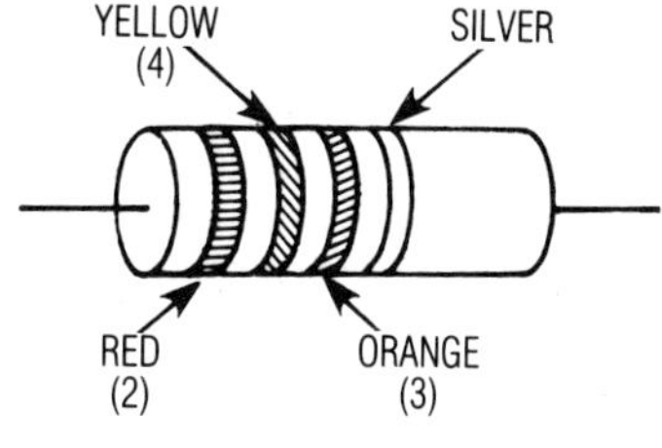

Fig. 2-2

Example 9
What is the resistance of a resistor which has rings red, orange, yellow, and silver?

1. Use Fig. 2-1 to refresh your memory and find the values of each color.
2. Red means 2.
3. Orange means 3.
4. Yellow means 4 zeros.
5. That makes the value of the resistor 230,000 ohms.
6. The fourth band means ±10%.
7. That means the value of the resistor's tolerance rating is 10% of 230,000, or 23,000 ohms.
8. Add the 23,000 to 230,000 to produce 253,000 for the + value of the resistor.
9. Subtract 23,000 from the value of the resistor and get 207,000 ohms for the − value.

Example 10
What is the resistance of a carbon-composition resistor that has rings of yellow, violet, and orange with a fourth band of gold?

1. Follow the same procedure as in Example 1.
2. The values of the colors are:

   a. yellow 4
   b. violet 7
   c. orange 3

3. This produces a resistance reading of 47,000 ohms.
4. The fourth band is gold. This equates to $\pm 5\%$.
6. 47,000 ohms $\times$ 0.05 = $\pm$2350 ohms.
7. 47,000 + 2350 = 49,350 ohms for the + value.
8. 47,000 − 2350 = 44,650 ohms for the − value.
9. The resistor is a 47K, or 47,000-ohm, resistor if it reads between 44,650 and 49,350 ohms.

If you've done any problems in color code, you probably noticed that the lowest value you can obtain from the color code as described so far is 10 ohms. That means brown, black, and black for the three bands, respectively. Now that the technology has made it possible to make resistors with smaller values, it takes a modification of the code to be able to color-code them to read less than 10 ohms.

By using gold and silver as the *third* band, it is possible to obtain less than 10 ohms with the code. Gold is used as 0.1 multiplier and silver is used as a 0.01 multiplier. For example, if you have a resistor with brown, black, and gold for the three bands, the value of the resistor is 1 for brown and 0 for the black, and that 10 is divided by 10 or multiplied by 0.1 to produce 1 ohm. If you want the lowest possible value obtainable with the code, you use brown, black, and silver. The brown is 1, the black ring is 0, and the silver means you divide bv 100 or multiply by 0.01 to produce 0.1 ohms. Hence, it is possible to obtain a tenth of an ohm by using the color code. The *fourth* band would still be gold or silver to represent the tolerance range of the resistor.

Example 11

What is the value of a resistor with rings of red, violet, and silver?

1. Check the table to obtain the values of the colors.
2. The color code reads as follows:
   a. red 2
   b. violet 7
   c. silver 0.01 (or divide by 100)
3. The 27 is divided by 100 to produce 0.27 ohms.
4. Or the 27 is multiplied by 0.01 to produce 0.27 ohms.

Example 12

What is the value of a resistor with rings of yellow, orange, and gold?

1. Check your color code to refresh your memory.
2. The color code values are:
   a. yellow 4
   b. orange 3
   c. gold as a third band means divide by 10 or multiply by 0.1.
3. The 43 is divided by 10 to produce 4.3 ohms.
4. Or you can multiply the 43 by 0.1 to produce 4.3 ohms.

PROBLEMS

1. What is the resistance *range* of a resistor that reads red, red, red, silver?
2. What is the resistance *range* of a resistor that reads yellow, orange, orange, gold?
3. What is the + *tolerance* of a resistor that reads red, black, red, silver?
4. What is the − *tolerance* of a resistor that reads red, yellow, green, gold?
5. What is the value of a resistor with rings of brown, green, gold?
6. What is the value of a resistor with rings of orange, violet, silver, silver?

OHM'S LAW

Georg Simon Ohm (1787–1854) was an early experimenter in electrical theory and originated Ohm's law, which we use so much today. His was a mathematical analysis of any circuit. He found that the voltage, current, and resistance in any circuit had a definite mathematical relationship to each other. Specifically, he found that the current in any circuit was equal to the voltage divided by the resistance.

Resistance is that opposition to the flow of electrons in a circuit or any device. It is measured in ohms (Ω), named to honor Georg Simon Ohm. Ohm also found that the resistance in any object depends on four things:

1. The composition of the material.
2. The temperature of the material.
3. The length of the material.
4. The cross-sectional area of the material.

As you can see, the temperature can affect the resistance of an object. Since all objects have resistance, it is necessary to apply a force to cause the electrons to move in a definite direction. This movement of electrons is called *current flow*. The current flow, then, is controlled by the amount of resistance and the voltage, or *electrical pressure*, applied to the object by the source of potential.

Circuits utilize the various voltages or electrical pressures to cause electrons to flow in a desired path or direction. Voltage is generated by a number of methods. A battery can be used to generate a difference of potential from one terminal to the other using chemical means. A generator uses magnetism to cause the electrons to move in a copper conductor. The application of heat can cause electrons to move as well, as can application of light to some surfaces.

The *electromotive force* (*emf*) is generated by any of a number of methods. It serves one purpose, to cause electrons to flow in a substance that contains resistance. Once a path for electrons is established for them to flow from one terminal (excess) to the other (deficiency), a difference of electrical charge can cause them to move along the wire or path through the resistance and back to the deficiency side of the battery or generator.

The limiting factor for electron flow (current) is the resistance of the circuit. A number of methods are used to make resistors with specific amounts of opposition or resistance.

In order to apply the Ohm's law relationship to circuits it is necessary to label the parts of the circuit so that they can be represented in a formula or equation. Electromotive force is called voltage and is represented in Ohm's law formulas by an E. Current flow is represented by the letter I to stand for the intensity of electron flow in the circuit. R stands for resistance. So, if we are to represent Ohm's law in a formula, it can be expressed in the following forms:

$$E = I \times R$$

$$I = \frac{E}{R}$$

$$R = \frac{E}{I}$$

It may be easier to remember the formulas by checking Fig. 2-3, below. Let your finger cover the letter representing the value you are looking for, and the remaining two letters will show you the relationship between them by being situated either one on top of the other (division) or one next to the other (multiplication).

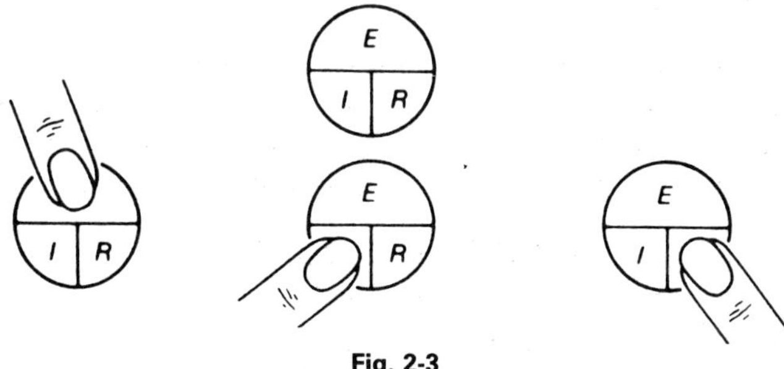

Fig. 2-3

Example 13

A resistor has a voltage of 12 volts applied to its 4 ohms of resistance. What is the current that will flow through the resistor?

1. Determine the unknown: I, or current, is the unknown; therefore the formula to use is:

$$I = \frac{E}{R}$$

2. Substitute the voltage and resistance in the formula. $R = 4$ ohms and $E = 12$ volts. So, $I = \frac{12}{4}$ or 3 amperes.

3. The voltage is divided by the resistance to produce a current of 3 amperes. Therefore, $I = 3$ A.

Example 14

How much voltage does it take to force 3 amperes of current through a resistor of 10 ohms?

1. Determine the unknown: E, the emf or voltage, is unknown; therefore the formula to use is:

$$E = I \times R$$

2. Substitute the current and the resistance in the formula. $I = 3$ amperes and $R = 10$ ohms. So, $E = 3 \times 10$ or 30 volts.
3. The current is multiplied by the resistance to produce a voltage of 30 volts. Therefore, $E = 30$ volts.

Example 15

What is the resistance of a circuit that has a voltage source of 50 volts pushing 5 amperes through a resistor?

1. Determine the unknown: R is the unknown; therefore the formula to use is:

$$R = \frac{E}{I}$$

2. Substitute the voltage and current in the formula. $E = 50$ volts and $I = 5$ amperes. So, $R = \frac{50}{5}$ or 10 ohms.
3. The resistance is found by dividing the voltage by the current. Therefore the resistance is 10 ohms.

PROBLEMS

1. What voltage is needed to force 5 amperes through 5 ohms?
2. What voltage is needed to force 2 amperes through 10 ohms?
3. What voltage is needed to force 1 ampere through 100 ohms?

4. How much of an emf is needed to push 3 amperes through a 100-ohm resistor?
5. How much of an emf is needed to push 2 amperes through a 47 ohm resistor?
6. What is the current through a 10-ohm resistor if the voltage of the battery is 10 volts?
7. What is the current through a 100-ohm resistor if the voltage of the battery is 10 volts?
8. What is the current through a 1000-ohm resistor if the voltage of the battery is 10 volts?
9. What is the current through a 470-ohm resistor if the voltage supply is putting out 47 volts?
10. What is the current through a 4700-ohm resistor if the voltage supply is putting out 47 volts?
11. What is the resistance of a circuit if the current is 2 amperes and the battery is putting out 20 volts?
12. What is the size of a resistor which has 1 ampere through it and a battery of 24 volts connected to it?
13. What is the resistance of a resistor if it has 0.25 amperes pushed through it by a 250-volt battery?
14. What is the size of the resistor which has 0.025 amperes pushed through it by a 25-volt battery?
15. What is the current through a resistor of 4700 ohms and a voltage of 4.7 volts?

CHAPTER 3

Circuits

POWER IN RESISTIVE CIRCUITS

Power is measured in watts. The watt was named for James Watt (1736–1819), who lived in Scotland and did experimental work in many areas, but is most generally known for his work with the steam engine. In electrical terms, the watt is found by multiplying the voltage times the current in a circuit or resistor. In a resistor, the power dissipated by the surface of the resistor in terms of heat is measured in watts. The wattage rating of resistors is determined by the physical size of the resistors. The larger the resistor, the greater its ability to dissipate heat to the surrounding medium.

The term used to indicate power in a formula is P for power. Power is measured in watts (W). Sometimes the term *kilowatt* (kW) is used. *Kilo* means 1,000, so the kilowatt is 1,000 watts.

Power in watts (P) = voltage $(E) \times$ amperes (I), or

$$P = E \times I$$

Therefore, if $P = E \times I$ and E is equal to $I \times R$, then you can substitute and produce $P = I \times R \times I$ or $P = I^2R$. Therefore, if $P = E \times I$ and I is equal to E/R, then you can substitute and obtain $P = E \times E/R$, or $P = \frac{E^2}{R}$.

This produces three formulas you can use to calculate the power if two other factors are known.

$$P = E \times I$$

$$P=\frac{E^2}{R}$$
$$P=I^2R$$

You may also find the current if you know the voltage and power:

$$I=\frac{P}{E}$$

Keep in mind that the expression I^2R is often used in electronics literature to mean power dissipation or power loss $(P=I^2R)$.

Example 1
What is the power consumed by a circuit with a current of 10 amperes and with 100 volts applied?

1. Determine the formula needed by checking what factors are known and which are unknown.
2. In this case you know the voltage and the current; therefore you use $P=E\times I$.
3. Substitute the values for E and I to get $P=100\times 10$.
4. Solve the equation and you get $10\times 100=1{,}000$ W.

Example 2
What is the power consumed by a resistor with 10 ohms of resistance and has a voltage drop of 100 volts across it?

1. Determine the formula needed by checking the known and unknown factors.
2. In this case you know the resistance and the voltage, so you need $P=\frac{E^2}{R}$.
3. Substitute the values for E and R to get 100 squared divided by 10, expressed by:

$$P=\frac{100^2}{10}$$

4. Solve the equation and you get $100\times 100=10{,}000\div 10$. The answer is 1000 W.

Example 3

What is the power loss across a resistance of 470 ohms when 100 milliamperes flow through it? Remember, a milliampere is 0.001 ampere.

1. Determine the formula needed by checking the known and unknown factors.
2. In this case you know the resistance and the current, so you need $P = I^2R$.
3. Substitute the values for I and R and you get 0.1 ampere $\times$ 0.1 ampere $\times$ 470.
4. Solve the equation and you get 4.7 W.

PROBLEMS

1. What is the power consumed by a resistor with 100 ohms of resistance and a voltage drop of 100 volts?
2. What is the power consumed by a resistor with 20 ohms of resistance and a voltage drop of 100 volts?
3. What is the power consumed by a resistor with 470,000 ohms of resistance and a voltage drop of 100 volts?
4. What is the power consumed by a resistor with 330,000 ohms of resistance and a voltage drop of 3 volts?
5. What is the power consumed by a resistor with 500 ohms of resistance and a voltage drop of 500 volts?
6. What is the power consumed by a resistor of 10 ohms of resistance and that has a current of 2 amperes through it?
7. What is the power consumed by a resistor of 20 ohms of resistance that has a current of 2 amperes through it?
8. What is the power consumed by a resistance of 1000 ohms and a current of 500 milliamperes (or 0.05A)?
9. What is the power consumed by a resistor of 200 ohms and a current of 50 milliamperes (or 0.05A)?
10. What is the power consumed by a resistor of 470k ohms and a current flow of 10 milliamperes (0.01A)?

11. What is the power consumed by a resistor if the voltage is 100 and the current is 1 ampere?
12. A circuit with 120 volts applied has a current of 15 amperes. What is the power consumed by the circuit?
13. A circuit with 240 volts applied has a current flow of 15 amperes. What is the power consumed?
14. A circuit with 500 volts applied has a current flow of only 750 milliamperes (or 0.75A). What is the power consumed?
15. A circuit with 1,000 volts applied is drawing 2 amperes. What is the power consumed by this circuit?

SERIES CIRCUITS

Series circuits consist of two or more resistors, capacitors, or inductors connected one after the other. In this section we will look primarily at resistors. Capacitors and inductors will be covered in a later chapter.

Series circuits are made up of two resistors connected end to end, as shown in Fig. 3-1. Current flow is from A to B or through both resistors. A series circuit has only *one* path for current flow from the excess-electrons terminal of the battery to the deficiency terminal of the power source. Therefore R_1 and R_2 both have the same amount of current flow. The power supply or battery supplies the electrical pressure (voltage) to force the electrons through the resistances of the resistors.

Three rules of series circuits aid in the solution of series circuit problems. They deal with the three factors found in any circuit—voltage, current, and resistance.

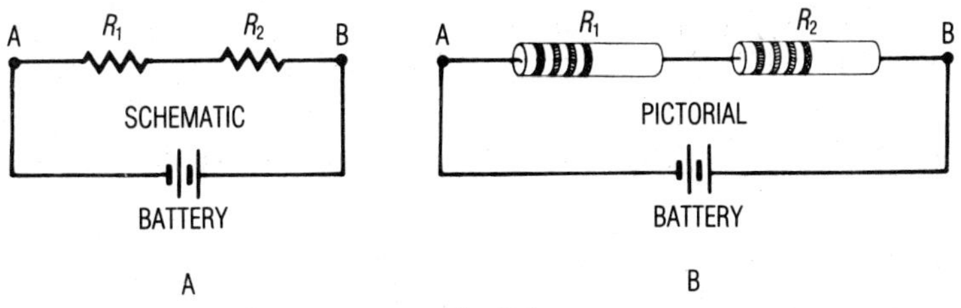

Fig. 3-1

Rules of Series Circuits

1. Current in all resistors is the *same* as the total current.

$$I_T = I_{R_1} = I_{R_2} = I_{R_3}$$

2. Voltage *divides* according to the resistance of each individual resistor. Or, the sum of the individual voltage drops across the resistors equals the applied voltage.

$$E_A = E_{R_1} + E_{R_2} + E_{R_3} + \ldots$$

3. Total resistance is found by *adding* the individual resistances.

$$R_T = R_1 + R_2 + R_3 + \ldots$$

Example 4

What is the total resistance of four resistors connected in series if the resistors have resistances of 10 ohms, 20 ohms, 30 ohms and 40 ohms?

1. Determine what is to be found. In this case it is the total resistance.
2. Note the formula for finding total resistance in a series circuit is:

$$R_T = R_1 + R_2 + R_3 + R_4$$

3. Substitute the resistances for the letters:

$$R_T = 10 + 20 + 30 + 40$$

4. Sum up the resistances to produce 100 ohms.

Keep in mind that voltage drop refers to the voltage *across* an individual resistor.

Example 5

What is the total resistance of the circuit shown in Fig. 3-2?

1. Determine what is to be found. In this case it is the total resistance.
2. Note the formula for finding the total resistance in a series circuit is:

$$R_T = R_1 + R_2 + R_3$$

3. Substitute the resistances for the letters in the formula:

$$R_T = 20 + 30 + 50$$

4. Sum up the resistances to produce 100 ohms.

Using Ohm's law, it is possible to determine the voltage drop across each resistor if the total current is given.

Example 6

What is the voltage drop across each resistor in Fig. 3-2 if the total current is 2 amperes?

1. Determine what you know at this point. Check the circuit to see what the resistor values are. They are 20, 30, and 50 ohms.
2. Check what the current is. Current is given as 2 amperes total.
3. Check the rules of the series circuit for current and you find that the total current is equal to the current through each resistor in series. That means, then, that the current through each resistor is 2 amperes.
4. Therefore you know the current through each resistor and you know the resistance. The formula (Ohm's law) for finding the voltage drop across each resistor is simply $E = I \times R$. So,

    a. $E_{R_1} = I \times R_1$ or $E_{R_1} = 2 \times 20$ or 40 volts
    b. $E_{R_2} = I \times R_2$ or $E_{R_2} = 2 \times 30$ or 60 volts
    c. $E_{R_3} = I \times R_3$ or $E_{R_3} = 2 \times 50$ or 100 volts

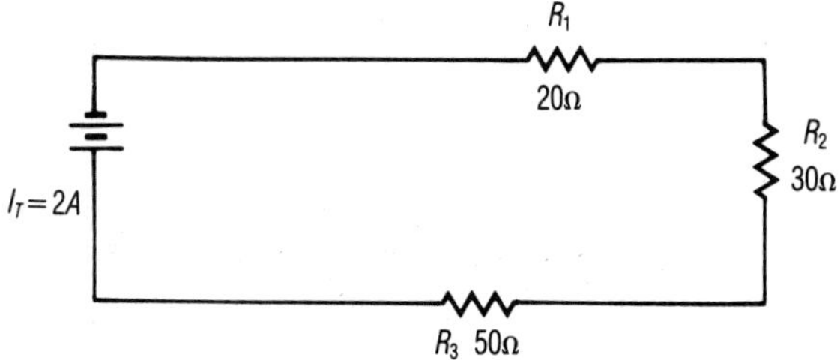

Fig. 3-2

Example 7

What is the voltage drop and the current through each resistor in Fig. 3-3?

1. Determine what is given or known. You know the resistance of each resistor and you know the total current to be 3 amperes.
2. Another thing you know is that the current in a series circuit is the *same* in every resistor. The total current is also the current through each resistor. That means each resistor has 3 amperes through it.
3. By using Ohm's law you can find the individual voltage drops, or $E = I \times R$.

   a. $E_{R_1} = 3 \times 30$ or 90 volts
   b. $E_{R_2} = 3 \times 40$ or 120 volts
   c. $E_{R_3} = 3 \times 50$ or 150 volts
   d. $E_{R_4} = 3 \times 60$ or 180 volts

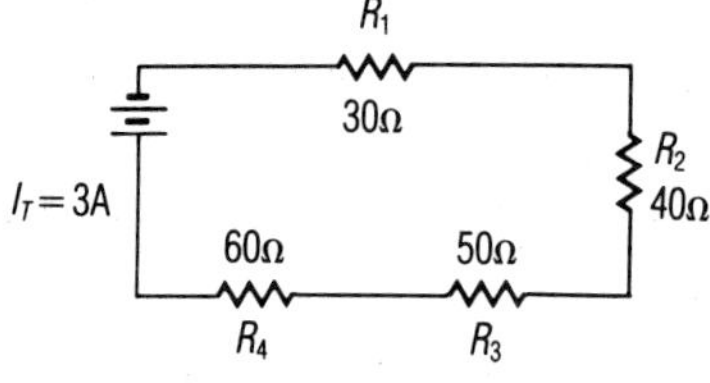

Fig. 3-3

4. To check your answer you can add the voltages to produce $90 + 120 + 150 + 180 = 540$ volts for the applied voltage.
5. You already know the total current to be 3 amperes. Using Ohm's law you can determine the total resistance by:

$$R_T = \frac{E_A}{I_T} \quad \text{or} \quad R_T = \frac{540}{3} \text{ or 180 ohms}$$

> E_A is the symbol used to indicate Applied Voltage (sometimes called Total Voltage).

6. Double-check your answer by adding the individual resistances to see if they produce 180 ohms. So, $30 + 40 + 50 + 60 = 180$, and your calculations are correct.

PROBLEMS

1. Two resistors of 10 ohms each are connected in series. What is the total resistance?
2. Three resistors of 10 ohms each are connected in series. What is the total resistance?
3. Three resistors of 10, 20, and 40 ohms are connected in series. What is the total resistance?
4. What is the total resistance of five resistors that have the following resistances: 5 ohms, 10 ohms, 100 ohms, 5,000 ohms, and 15,000 ohms?
5. If you have three resistors of 10 ohms, 20 ohms, and 30 ohms connected in series and apply a voltage of 120 volts to the circuit, what would be the current through each resistor?
6. There are four resistors connected in series. Each resistor is the same size (470 ohms). What is the voltage drop across each if a current is detected at 200 milliamperes (0.2A) in one of them?
7. What is the current through four resistors of 470 ohms, 670 ohms, 82 ohms, and 278 ohms if the voltage supplied by the battery is 24 volts?
8. What is the current through a series circuit if the total current is 2 amperes?
9. What is the voltage applied by a battery to a series of resistors each with a current of 2 amperes and resistances of 15 ohms, 40 ohms, 35 ohms, and 75 ohms?
10. What is the resistance of the third resistor in a series circuit if the total resistance is 100 ohms and the other two resistors have 40 ohms and 20 ohms as their values?
11. What is the current through a series combination if the applied voltage is 100 and the individual resistors are 20 ohms, 30 ohms, 60 ohms, and 90 ohms?
12. The total resistance of three resistors is 500 ohms. Two of the resistors each have 100 ohms of resistance. What is the resistance of the third resistor?
13. You have a string of ten lamps each with a rating of 12 volts.

What happens to the voltage across each resistor if one burns out and you short across it to cause the others to glow?

14. What happens to a circuit's resistance if the three resistors in series happen to short causing an excess current that makes one of the resistors open?
15. What is the total resistance of a circuit you are measuring if one of the series resistances opens?
16. Find the total resistance of the following series resistors:

	No. of Resistors	Values of Resistors	Total Resistance
a.	4	10, 20, 30, 40	
b.	5	20, 30, 40, 50, 60	
c.	6	30, 40, 50, 60, 70, 80	
d.	7	40, 50, 60, 70, 80, 90, 100	
e.	8	50, 60, 70, 80, 90, 100, 110, 120	

PARALLEL CIRCUITS

The parallel circuit is different from the series as much as it has two or more resistors connected across the power supply. See Fig. 3-4. The resistors are across the power source, so the applied voltage and the voltage across each resistor are the same.

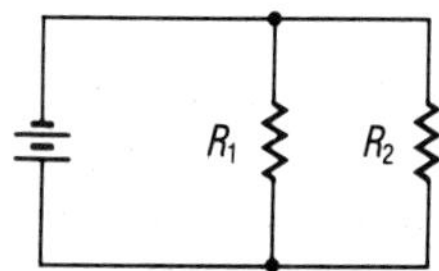

Fig. 3-4

The total current in a parallel circuit can be found by adding the individual branch or loop currents. Or:

$$I_T = I_{R_1} + I_{R_2} + I_{R_3} + \ldots$$

The total resistance becomes less as the number of resistors are added to the circuit. This calls for some special treatment in the way of formulas. The law that controls this type of situation is: *The equivalent resistance of a parallel resistance network is the reciprocal of the sum of the reciprocals of the individual resistors.*

When stated as a formula, this becomes:

$$R_T = \frac{1}{\frac{1}{R_1}+\frac{1}{R_2}+\frac{1}{R_3}+\frac{1}{R_4}+\frac{1}{R_5}+\frac{1}{R_6}+\ldots}$$

Another way to put it is:

$$\frac{1}{R_T} = \frac{1}{R_1}+\frac{1}{R_2}+\frac{1}{R_3}+\ldots$$

If you have only *two* resistors in parallel, you can use:

$$R_T = \frac{R_1 \times R_2}{R_1 + R_2}$$

Example 8

What is the equivalent resistance of a 10-ohm and a 20-ohm resistor when the two are connected in parallel?

1. Determine the unknown. In this case the total resistance is unknown.
2. Since there are two resistors here, you have a choice of formulas: You can use the product divided by the sum or the reciprocal formula. Let's use the product divided by the sum and see how easy it is first.
3. $$R_T = \frac{R_1 \times R_2}{R_1 + R_2}$$
4. Substitute the values of the resistors in the formula.
5. $$R_T = \frac{10 \times 20}{10 + 20} = \frac{200}{30} = 6.667 \text{ ohms}$$
6. Note: *In parallel circuits the total resistance is always less than the smallest resistor*. This will give you a quick check on whether you are close to being correct or way off in your answer.

Example 9

What is the equivalent resistance of a 10-ohm and a 20-ohm resistor when the two are connected in parallel?

1. This time we will use the reciprocal formula and see if we get the same answer as with the product-and-sum formula.
2. $$\frac{1}{R_T}=\frac{1}{R_1}+\frac{1}{R_2}$$
3. Substitute the resistor values in the formula to get:
$$\frac{1}{R_T}=\frac{1}{10}+\frac{1}{20}$$
4. Solve by finding the common denominator. The lowest
   a. $$\frac{1}{R_T}=\frac{2+1}{20}$$
   b. $$\frac{1}{R_T}=\frac{3}{20}$$
   c. Invert to get $R_T=\frac{20}{3}=6.667$ ohms.
   d. Note that the answer is the same as we got using the other method. Therefore, it must work either way.

Example 10

What is the total resistance of four resistors connected in parallel if their individual resistances are 10, 20, 30, and 60 ohms?

1. Determine the formula to use. In this case there are more than two resistors, so it is necessary to use the reciprocal formula.
2. $$\frac{1}{R_T}=\frac{1}{R_1}+\frac{1}{R_2}+\frac{1}{R_3}+\frac{1}{R_4}$$
3. Substitute the values of the resistors in the formula:
$$\frac{1}{R_T}=\frac{1}{10}+\frac{1}{20}+\frac{1}{30}+\frac{1}{60}$$
4. Determine the common denominator. In this case it is 60.
5. Perform the indicated operations to produce:
   a. $$\frac{1}{R_T}=\frac{6+3+2+1}{60}$$

b.
$$\frac{1}{R_T}=\frac{12}{60}$$

c.
$$R_T=\frac{60}{12}$$

d. Solve to produce 5 ohms for total resistance. *Note*: Once again the 5 ohms is smaller than the smallest resistor.

If you have a calculator with a reciprocal key, it is much easier to obtain the result in a problem of this type. Take the reciprocal of each and add them. Then take the reciprocal of the sum of the reciprocals.

Example 11

What is the total current for the circuit shown in Fig. 3-5?

1. Determine the formula to be used. In this case you already know the total current in a parallel circuit is equal to the sum of the individual currents.
2. $I_T=I_{R_1}+I_{R_2}+I_{R_3}$
3. $I_T=3+2+4$
4. $I_T=9$ amperes.

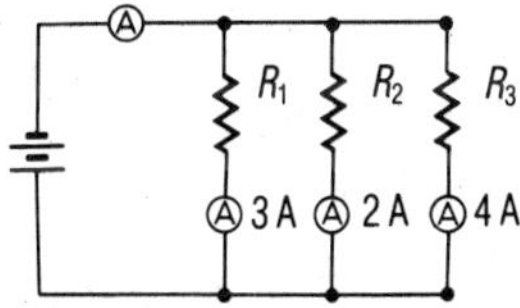

Fig. 3-5

Example 12 (Using Ohm's law to solve parallel circuits)

Ohm's law is very useful in finding unknowns in a parallel circuit. The combination of Ohm's law and the three rules for parallel circuits makes it possible to solve for some unknowns in a circuit that would otherwise be unobtainable.

What are the voltage, current, and resistance for the circuit shown in Fig. 3-6?

1. Determine the steps to be followed.
   a. $I_{R_1}=?$

b. $R_2 = ?$
c. $I_T = ?$
d. $R_T = ?$
e. $E_A = ?$

2. Substitute and solve.

a. $I_{R_1} = \dfrac{E}{R_1}$ or $I_{R_1} = \dfrac{20}{10} = 2$ amps

b. $R_2 = \dfrac{E}{I_{R_2}}$ or $R_2 = \dfrac{20}{4} = 5$ ohms

c. $I_T = I_{R_1} + I_{R_2}$ or $I_T = 2 + 4 = 6$ amps

d. $R_T = \dfrac{E_A}{I_T}$ or $R_T = \dfrac{20}{6} = 3.33$ ohms

3. To check your answer, use:

$$R_T = \frac{R_1 \times R_2}{R_1 + R_2} \quad \text{or} \quad R_T = \frac{10 \times 5}{10 + 5} = \frac{50}{15} = 3.33 \text{ ohms}$$

4. $E_A = 20$ V

5. Therefore, $E_{R_1} = 20$ V and $E_{R_2} = 20$ V since the resistors are in parallel.

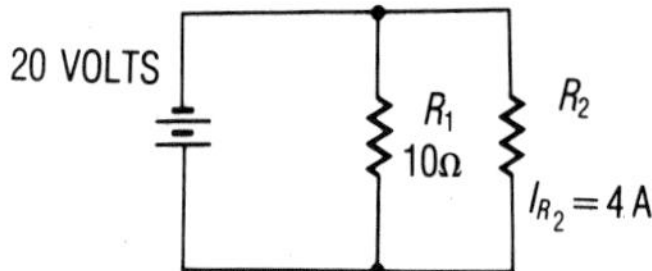

Fig. 3-6

PROBLEMS

1. What is the equivalent or total resistance of two resistors in parallel if each has 50 ohms resistance?
2. What is the total resistance of two resistors of 50 ohms and 60 ohms in parallel?
3. What is the total resistance of three resistors of, respectively, 100, 1,000, and 10,000 ohms connected in parallel?
4. A 470,000-ohm resistor is connected in parallel with a 470-ohm resistor. What is the total resistance?

5. Three resistors of 1,000 ohms each are connected in parallel. What is the total resistance?
6. If two resistors of 50 and 100 ohms are connected in parallel and have 100 volts applied to the circuit, what is the total current in the circuit?
7. Three resistors of 60, 100, and 120 ohms are connected across a voltage source of 240 volts. What is the total current in the circuit?
8. How many 1000-ohm resistors have to be connected in parallel to produce 250 ohms equivalent resistance?
9. If you had ten 10-ohm resistors, what would the ohmmeter measure if they were connected in parallel?
10. How many lamps can you connect in parallel across a 120-volt power supply that can produce 15 amperes if each lamp has 64 ohms resistance?
11. If you have a transistor battery of 9 volts connected across a parallel circuit with 470,000, 330,000 and 4.7 megohms, what would be the current drain?
12. What is the voltage drop across resistors in parallel if their total resistance is 20 ohms and the total current is 2 amperes?
13. What is the current through each of the following resistors if the power supply is 100 volts? The resistors are 100, 200, and 400 ohms.
14. What is the total resistance of 5 resistors connected in parallel if each has a resistance of 100 ohms?
15. You wish to make a resistor combination that will safely handle 5 watts of power, but you only have ½-watt resistors. If you have a need for 1,000 ohms at 5 watts, but have only 10,000-ohm ½-watt resistors, how many would you have to put in parallel to obtain the wattage and resistance needed?

SERIES-PARALLEL CIRCUITS (RESISTIVE ONLY)

Series-parallel circuits consist of a combination of a parallel and a series circuit. It takes a minimum of three resistors to produce a series-parallel combination. Two have to be in parallel and the other in series with the parallel group.

In most instances you will find this type of circuit used in electronics circuits. The parallel circuit is used almost exclusively in home wiring and appliance applications. Series circuits have very little practical use except in control applications. So, a combination of the two with the resulting advantages of each combined makes for a unique type of circuit with many applications.

The combination of resistors in series-parallel are referred to as a network. This network has to be *reduced* to a series *equivalent* in order for the total resistance to be found. That is where most of the difficulty in solving this type of circuit lies. Therefore, we will spend our time here showing how the circuit is reduced to a series equivalent so that the resistors can be simply added to obtain the total resistance.

Example 13

Find the total resistance of the circuit shown in Fig. 3-7.

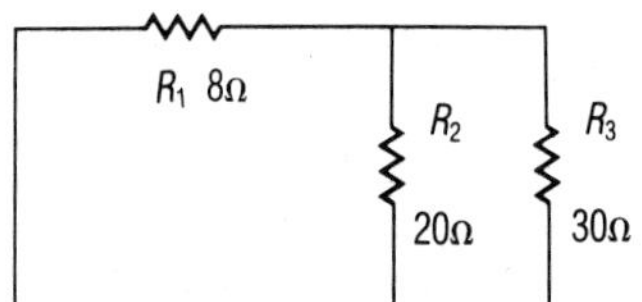

Fig. 3-7

1. Determine the procedure to be used.
2. Note that R_2 and R_3 are in parallel. Reduce this parallel combination to one equivalent resistor.
3. $$R_a = \frac{R_2 \times R_3}{R_2 + R_3}$$
4. Substitute values into the formula.
5. $$R_a = \frac{20 \times 30}{20 + 30}$$
6. Solve the equation: $20 \times 30 = 600$ and $20 + 30 = 50$. Divide the 600 by 50 to produce 12 ohms.

7. This means the circuit has been reduced to that shown in Fig. 3-8. Since this is now a series circuit, the total resistance can be found by adding the two resistances.
8. Add 8 and 12 to get 20 ohms for the total resistance of the series-parallel network.

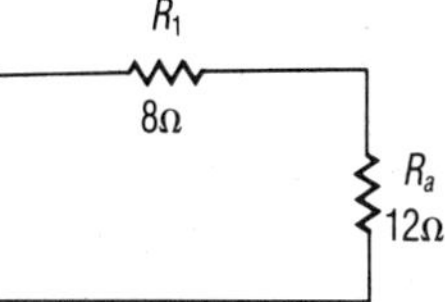

Fig. 3-8

Example 14

Find the total resistance in the circuit shown in Fig. 3-9.

1. Determine the method to be used to reduce it to a series circuit.
2. First, note that the two end resistors, R_3 and R_4, are in series. Add these two to get R_a and 60 ohms.
3. Now redraw the circuit so it looks like Fig. 3-10. Note how R_a is in parallel with R_2. Reduce this to a series equivalent by using the product divided by the sum of the two resistors, or:

$$R_b = \frac{R_2 \times R_a}{R_2 + R_a}$$

4. Substitute the resistances into the formula and get:

$$R_b = \frac{40 \times 60}{40 + 60}$$

5. Solve the equation to get $40 \times 60 = 2400$ and $40 + 60 = 100$. 2400 divided by $100 = 24$ ohms. That means R_b is now 24 ohms and in series with R_1.
6. Since R_1 and R_b are in series, as shown in Fig. 3-11, they can be added to produce the total resistance of $6 + 24 = 30$ ohms.
7. The total resistance of the series-parallel network shown in Fig. 3-9 is 30 ohms.

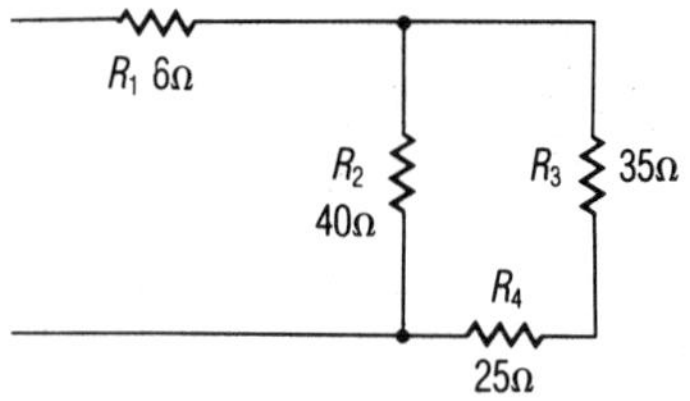

Fig. 3-9

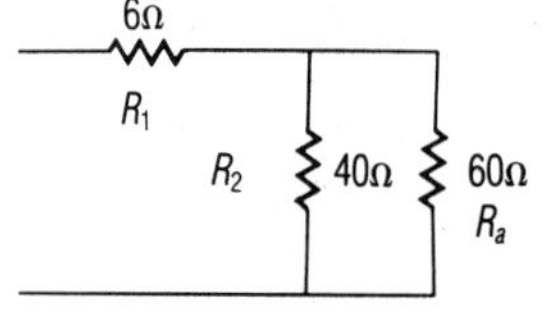

Fig. 3-10

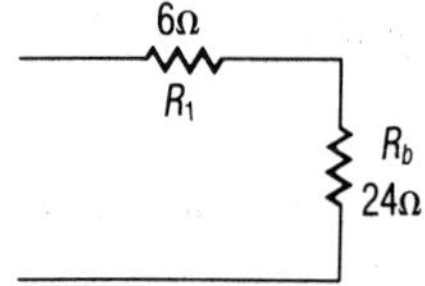

Fig. 3-11

Example 15

What is the total resistance of the series-parallel network shown in Fig. 3-12?

1. Determine the procedure to use. Note that R_5 and R_6 are in series. Add them to get $R_a = 30$ ohms and redraw the circuit to Fig. 3-13.
2. Observe that a parallel configuration exists with R_3 and R_4. Since there are only two in parallel, use the product divided by the sum formula, or:

$$R_b = \frac{R_3 \times R_4}{R_3 + R_4}$$

3. Substitute resistor values in the formula to produce:

$$R_b = \frac{20 \times 30}{20 + 30} = \frac{600}{50} = 12 \text{ ohms}$$

4. Since R_b is 12 ohms, it is also in series with R_2 with 8 ohms,

so the total resistance between points A and B (Fig. 3-13) becomes 20 ohms (R_c) and the resulting circuit, when redrawn, resembles Fig. 3-14.

5. A parallel combination still exists with R_a and R_c. Reduce this to a single equivalent resistance by using the product over the sum formula, or:

$$R_d = \frac{R_c \times R_a}{R_c + R_a}$$

6. Substitute resistor values in the formula to produce:

$$R_d = \frac{20 \times 30}{20 + 30} = \frac{600}{50} = 12 \text{ ohms}$$

7. Since R_c and R_a reduce to a single equivalent resistance of 12 ohms, it is then added with R_1 and R_7 (see Fig. 3-15) to get the total resistance of the series-parallel network.
8. Add $R_1 + R_d + R_7$ since they are in series:

$$R_T = 10 + 12 + 8 = 30 \text{ ohms}$$

9. The whole network has been reduced to a single equivalent resistance of 30 ohms. If an ohmmeter were placed across the network, it would read 30 ohms.

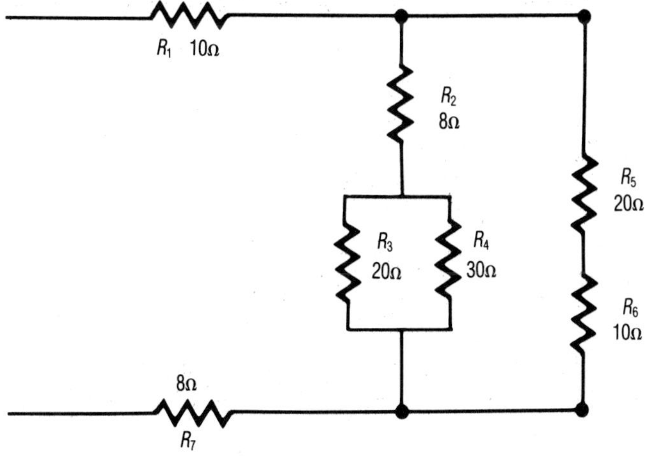

Fig. 3-12

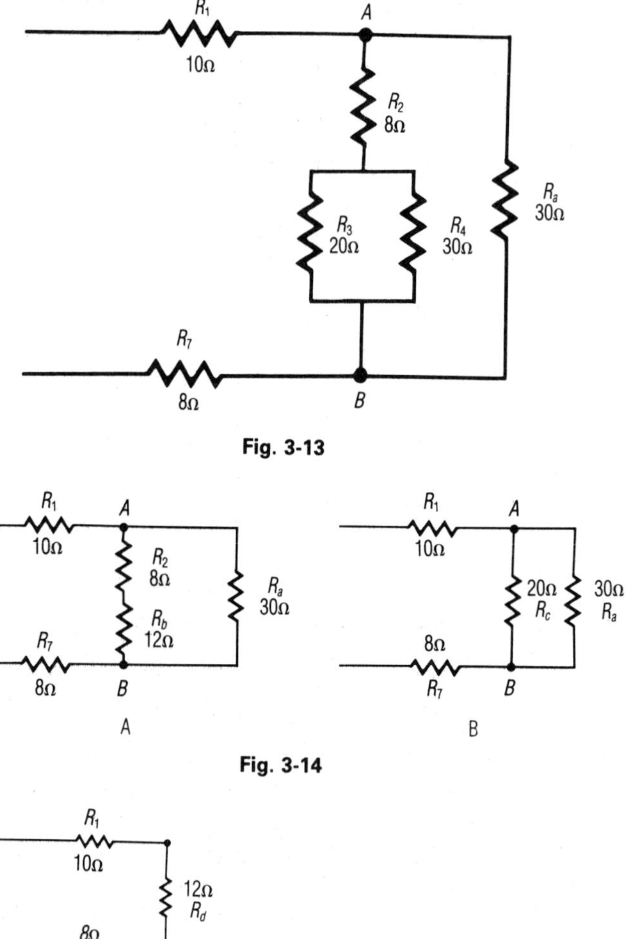

Fig. 3-13

Fig. 3-14

Fig. 3-15

PROBLEMS

1. Find the total resistance for the circuit in Fig. 3-16.
2. Find the total resistance for the circuit shown in Fig. 3-17.
3. Find the equivalent resistance of the circuit shown in Fig. 3-18.

4. Find the equivalent or total resistance of the circuit shown in Fig. 3-19.
5. Find the total resistance of the circuit shown in Fig. 3-20.

VOLTAGE, CURRENT, AND RESISTANCE IN A SERIES-PARALLEL CIRCUIT

Since the series-parallel circuit is common and its functioning can make the difference between an operational piece of equipment or non use of the device, it is important to take a close look at what the voltage distribution and current paths do in such a circuit. An understanding of the voltage distribution and the current in different parts of the circuit aid in troubleshooting the circuit and the possibility of putting the device or piece of equipment back in operation.

The circuits shown here will deal only with resistance. Resistive circuits are not particularly sensitive to whether the current is a.c. or d.c. However, this changes when a capacitor or inductor is introduced to an a.c. circuit. That will be handled later.

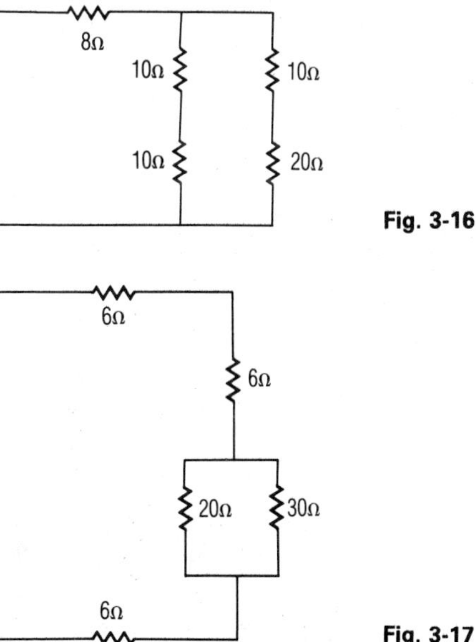

Fig. 3-16

Fig. 3-17

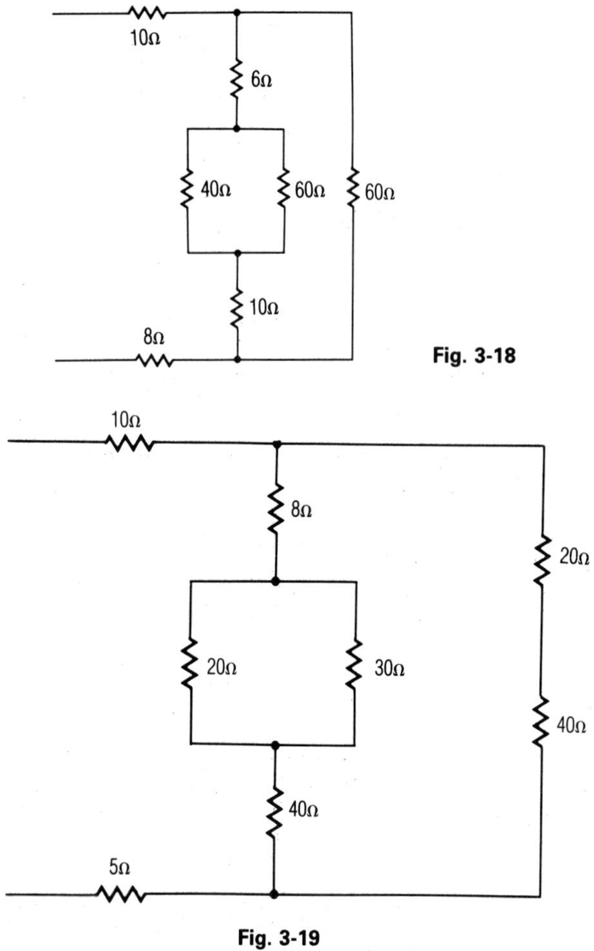

Fig. 3-18

Fig. 3-19

So far you know the three rules for a series circuit:

1. Current is the same in all parts of the circuit.
2. Voltage divides. You can obtain the applied voltage by adding the individual voltage drops.
3. Total resistance is found by adding the individual resistances.

The parallel circuit also has three rules which govern its current, voltage, and resistance:

1. Current divides. Total current can be found by adding the individual branch currents.
2. Voltage is the same across all resistors in parallel.
3. Total resistance can be found by using the reciprocal formula or the product divided by the sum in the case of only two resistors.

Additional Information on Parallel Resistors

If two resistors are in parallel and they are the same size, the total or equivalent resistance is half of the value of one. For instance, if you have two 10-ohm resistors, the equivalent resistance is 5 ohms.

If you have three resistors and they are in parallel and the same size, you can divide the value of one by the number 3 and obtain the total resistance. For instance, if you have three 300-ohm resistors, the total resistance is 100 ohms.

The same holds true for any number of equal resistors. If you have 100 1,000-ohm resistors in parallel, the total or equivalent resistance would be 10 ohms. All you do is divide the value of one of the resistors by the number of resistors, in this case 1,000 divided by 100 = 10 ohms. Try it and then use an ohmmeter to check to see

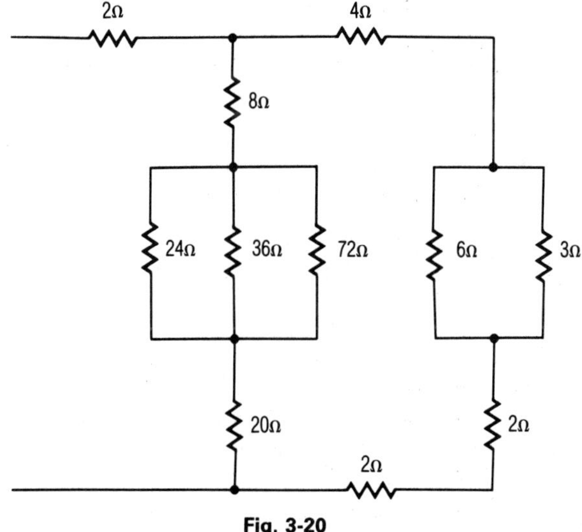

Fig. 3-20

if you are correct. It should work if the tolerance of the resistors are ±5%.

So far you have found the total resistance of the series-parallel circuit. Now it is time to look at the distribution of current and how it affects the voltage drop across each of the resistors in series-parallel. The first step, of course, is to reduce the resistors to one equivalent. This was explained in the previous unit. Now, let's take a couple of examples to see how the current is distributed in a series-parallel circuit.

Example 16

What is the voltage across each resistor and what is the current through each resistor in Figure 3-21?

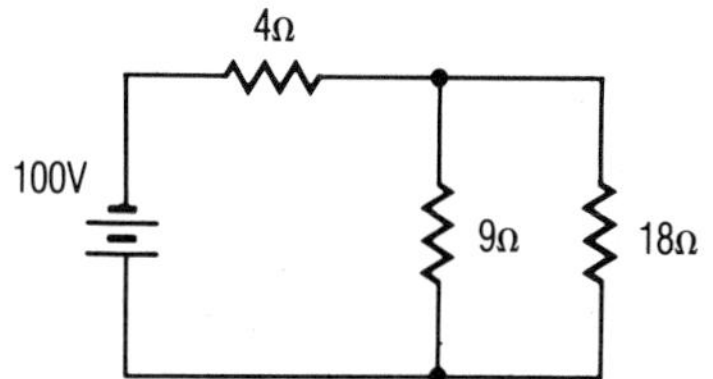

Fig. 3-21

1. You know the voltage as 100 volts, You also know the resistances.
2. Find the total resistance. In this case reduce the parallel loop by multiplying 9×18 and dividing it by $9 + 18$. This gives you a resistance of the parallel branch of 6 ohms. Add this 6 ohms to the series resistor of 4 ohms to get 10 ohms for total resistance.
3. If you have 100 volts applied and you have 10 ohms for total resistance, then the total current is 10 amperes. This is found by using Ohm's law: $I = \frac{E}{R}$.
4. The total current flows through the series resistor of 4 ohms. That means, using Ohm's law, that the 4 ohms times the 10 amperes produces 40 volts dropped across the series resistor.
5. If the total voltage is 100 and you just dropped 40, that leaves only 60 across the parallel branch.

6. With 60 volts across the 9 ohms it gives 6.667 amperes through the 9-ohm resistor.
7. Since the 9-ohm and 18-ohm resistors are in parallel, they have the same voltage across them.
8. Since the 9-ohm resistor had 60 volts across it, so does the 18-ohm resistor. So, divide the 60 by 18 to get the current through the 18-ohm resistor, or 3.33 amperes.
9. You can check to see if you are right by adding the currents through the parallel loops to see if they total the total current.
10. 6.67 + 3.33 equals 10 amperes or the total current. So you are right, because current flowing up to the junction point flows away from it. It simply divided at the junction of the two parallel resistors to recombine at the other end and flow back to the battery.

Example 17

What is the voltage drop and current through each of the resistors in Figure 3-22?

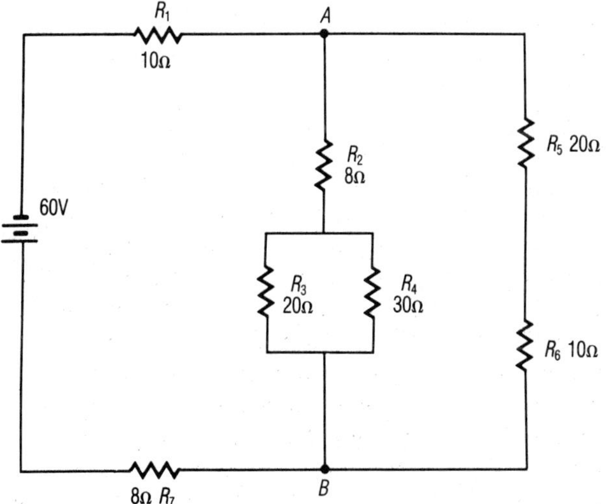

Fig. 3-22

1. Find the total resistance.
   a. Reduce the parallel combination of R_3 and R_4 to 12 ohms.
   b. That makes the 8 ohms of R_2 and the 12 ohms parallel equivalent in series, so that the total from *A* to *B* is 20 ohms.
   c. R_5 and R_6 are in series, so add them to get 30 ohms.
   d. That makes the *A* to *B* total now 20 ohms in parallel with 30 ohms.

$$\frac{20 \times 30}{20 + 30} = 12 \text{ ohms}$$

   e. Since *A* to *B* has now been reduced to 12 ohms, you can redraw the circuit so that the equivalent 12 ohms and the 10 ohms of R_1 and 8 ohms of R_7 add to it to make a series circuit equivalent of 30 ohms total. Therefore, the total circuit resistance is 30 ohms.
   f. Now that you have the total resistance you can use it to divide into the applied voltage 60 volts (Ohm's law).
   g. Total current is 2 amperes. That means R_1 and R_7 both have 2 amperes through them.

2. To find the voltage drop across R_1 multiply the resistance times the current, or 10×2, to give you 20 volts across R_1.
   a. To find the voltage drop across R_7 just multiply the 8 ohms times the 2 amperes to give 16 volts across R_7.
   b. From *A* to *B* is an equivalent of 12 ohms. This 12 ohms times the 2 amperes equals 24 volts. That means the full 60 volts is distributed around the loop, but you still don't know what the voltage is across each resistor. Next, look at *A* to *B* and you find that there are 24 volts from one point to the other. You can also see that there are 30 ohms in the right-side combination of resistances. This 30-ohm resistance has a voltage of 24 volts applied across it. That means Ohm's law can be used to obtain the current through the series resistors R_5 and R_6. 24 divided by 30 produces 0.8 ampere through the two resistors—they are in series, so each has the same current—which can be used to find the voltage across each one.
   c. The 0.8 ampere times the 20 ohms produces 16 volts across R_5.

d. The 0.8 ampere times the 10 ohms produces 8 volts across R_6.

e. Go back now to points A and B and see that R_2 has 1.2 amperes through it. How? Well, you know that the total current flowed through R_1 up to point A. It divided at point A. You already found that 0.8 amperes flowed through the R_5, R_6, path. That means the rest had to flow through R_2 and down.

f. Since 0.8 amperes went through R_5, R_6, you have 1.2 amperes through R_2. Take this 1.2 and multiply it times the resistance of 8 ohms to produce a voltage drop of 9.6 volts.

g. If 9.6 volts are dropped across R_2 and you had 24 volts from A to B, then R_3 will have $24 - 9.6$ or 14.4 volts across it. That means R_4 also has 14.4 volts across it since it is in parallel with R_3.

h. Divide the 14.4 volts by the resistance to produce the current through R_3. So, $I = \frac{14.4}{20}$ or 0.72 ampere.

i. Divide the 14.4 volts by the resistance to produce the current through R_4. So, $I = \frac{14.4}{30}$ or 0.48 ampere.

3. Now that you have all the currents you can check to see if you are right.

   a. $0.48 + 0.72 = 1.2$ amperes. That checks because 1.2 amperes flowed through R_2.

   b. Check again by adding the 1.2 amperes to the 0.8 amperes that flowed through R_5, R_6. This adds up to 2.0 amperes. This was the total current that will combine at point B and flow back to the other side of the battery. So, you had 2 amperes leaving and 2 amperes returning to the battery.

Now, let's see if you can do a few problems yourself.

PROBLEMS

1. Find the total resistance of the series-parallel combinations in Figure 3-23. A=____ B=____ C=____.

2. Find the current and voltage drop for each resistor shown in Figure 3-24.

$R_1 = 10\ \Omega$	$E_{R_1} =$	$I_{R_1} =$
$R_2 = 6\ \Omega$	$E_{R_2} =$	$I_{R_2} =$
$R_3 = 40\ \Omega$	$E_{R_3} =$	$I_{R_3} =$
$R_4 = 60\ \Omega$	$E_{R_4} =$	$I_{R_4} =$
$R_5 = 20\ \Omega$	$E_{R_5} =$	$I_{R_5} =$

KIRCHHOFF'S LAWS

Gustav Robert Kirchhoff was a German physicist better known for his pioneering work in spectroscopy than in electricity. He was a strong mathematician and got his education at the University of Konigsberg. In the mid-nineteenth century, he was familiar with Ohm's work and formulated the general rules that now bear his name. The rules are so simple, you wonder why someone didn't come up with them before.

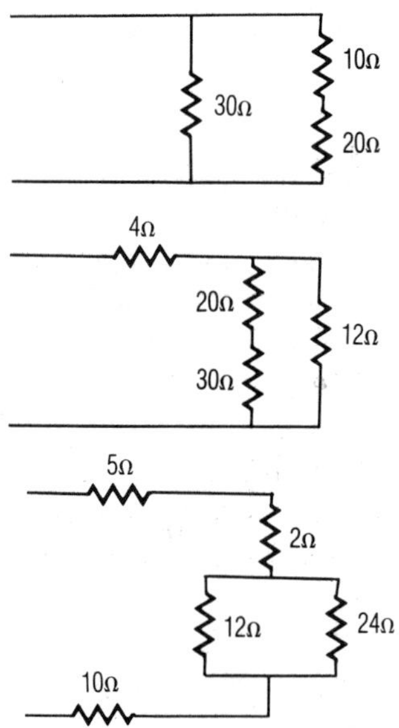

Fig. 3-23

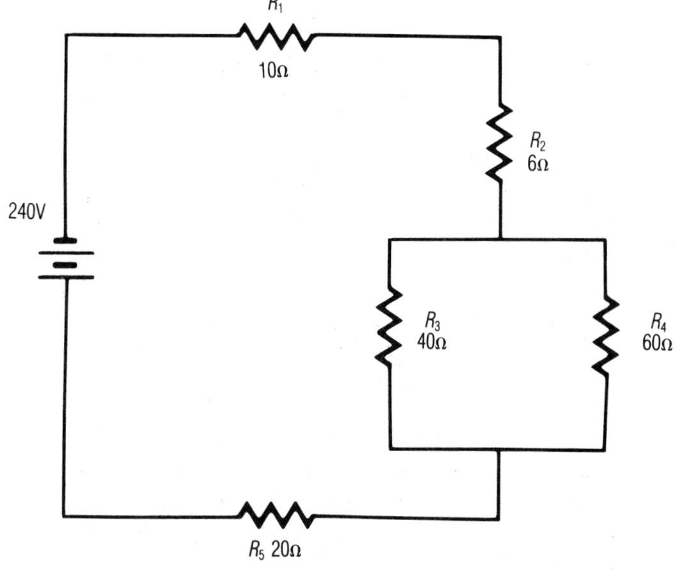

Fig. 3-24

Kirchhoff's First Law

This is now referred to as *Kirchhoff's point law.* This law simply states that at any point in a circuit the total current entering the point is equal to the total current leaving that point. The idea is to show that electrons do not pile up at any point in the circuit. See Fig. 3-25 for current dividing.

Kirchhoff's Second Law

This is now referred to as *Kirchhoff's loop law.* This law states simply that the sum (algebraic) of the voltage drops around any loop from

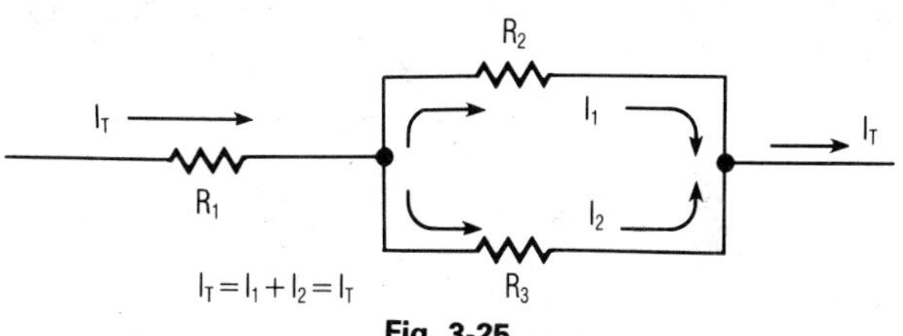

Fig. 3-25

one side of the battery to the other is equal to zero. The applied voltage is given a positive value and each of the individual voltage drops is assigned a negative value. The sum of the negative values will equal the positive source voltage, so the algebraic sum of the voltage drops around a loop is equal to zero. Take a look at Fig. 3-26 for the loops and the voltage drops around a complete loop.

Example 18

Determine the direction and the magnitude of the current through resistor R_3 of Fig. 3-27. Keep in mind that if the answer is negative, you have assumed the current flow is in the wrong direction.

1. Kirchhoff's first law states that the current flowing to the point flows away from that point.
2. Establish the current loops by inspection.
   I_1 goes from F to E, through R_3 to B, through R_1 to A, and through E_a to F. I_2 goes from D to E, through R_3 to B, through R_2 to C, and through E_b to D.
3. Substitute terms in the formulas:

$$E_a = (R_3)(I_1 + I_2) + (R_1)(I_1)$$
$$E_b = (R_3)(I_1 + I_2) + (R_2)(I_2)$$

4. Substitute values for one power source and simplify:

$$\begin{aligned} E_a\!:\ 12 &= 6{,}600(I_1 + I_2) + 9{,}400(I_1) \\ 12 &= 6{,}600I_1 + 6{,}600I_2 + 9{,}400I_1 \\ 12 &= 16{,}000I_1 + 6{,}600I_2 \end{aligned}$$

5. Do the same for the other power source (E_b):

$$\begin{aligned} E_b\!:\ 12 &= R_3(I_1 + I_2) + (R_2)(I_2) \\ 12 &= 6{,}600(I_1 + I_2) + 20{,}000I_2 \\ 12 &= 6{,}600I_1 + 6{,}600I_2 + 20{,}000I_2 \\ 12 &= 6{,}600I_1 + 26{,}600I_2 \end{aligned}$$

6. Solve for I_2 by simultaneous equations:

   E_a: $12 = 16{,}000I_1 + 6{,}600I_2$ (multiply this equation by 66)
   E_b: $12 = 6{,}600I_1 + 26{,}600I_2$ (multiply this equation by -160)

   This will result in two I_1's having the same value.

7. Multiply both sides of the equal sign in both equations and subtract E_b from E_a:

$$\begin{array}{lrl} E_a: & 792 = & \cancel{1{,}056{,}000}I_1 + 435{,}600I_2 \\ E_b: & -1920 = & -\cancel{1{,}056{,}000}I_1 - 4{,}256{,}000I_2 \\ \hline & -1128 = & -3{,}820{,}400I_2 \end{array}$$

8. To find I_2 substitute the value of I_2 in the E_a formula:

   a. $3{,}820{,}400I_2 = 1128$

   b. $I_2 = \dfrac{1128}{3{,}820{,}400}$

   c. $I_2 = 0.00029525704$ amperes

   d. E_a: $12 = 16{,}000I_1 + 6{,}600I_2$

$$12 = (16{,}000 \times I_1) + (6{,}600 \times 0.00029525704)$$
$$12 = 16{,}000I_1 + 1.948696471$$
$$16{,}000I_1 = 10.05130353$$
$$I_1 = \frac{10.05130353}{16{,}000}$$
$$I_1 = 0.0006282064706 \text{ amperes}$$

9. Determine the current through R_3:
   a. Current through $R_1 = I_1$, which is 0.0006282064706 amperes
   b. Current through $R_2 = I_2$, which is 0.0002952570411 amperes
   c. Current through $R_3 = I_1 + I_2$, or 0.0009234635117 amperes
10. *Note*: Since the value of R_3 is positive, the assumed direction of the current is correct. If the value were negative, the value would be correct but the direction would be wrong. Since both currents are positive, the assumed direction of current is correct. The current flows upward through R_3.
11. Voltage drop across $R_1 = 9{,}400 \times 6.282064706 \times 10^{-4} = 5.905140824$ volts.
12. Voltage drop across $R_2 = 20{,}000 \times 2.952570411 \times 10^{-4} = 5.905140822$ volts.
13. Voltage drop across $R_3 = 6{,}600 \times 9.234635117 \times 10^{-4} = 6.094859177$ volts.
14. Kirchhoff's second law states that the sum of the voltage

drops around a complete loop equals the applied voltage. That means:

$$\begin{array}{r} 5.905140822 \\ +\ 6.094859177 \\ \hline 12.000000000 \end{array}$$

Example 19

Determine the direction and magnitude of the current through resistor R_3 in Fig. 3-28. It is assumed that the current through R_3 is what the arrows indicate.

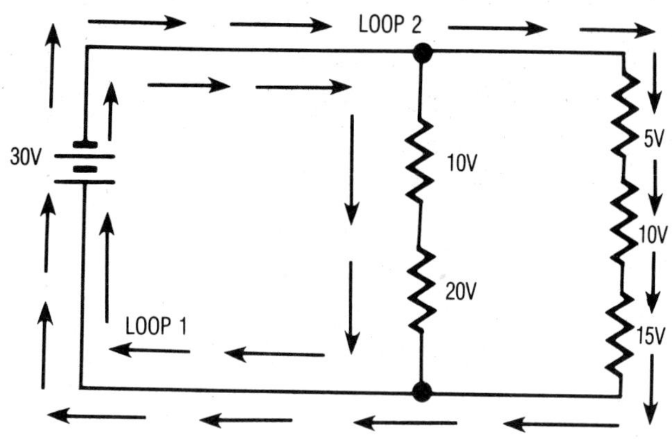

Fig. 3-26

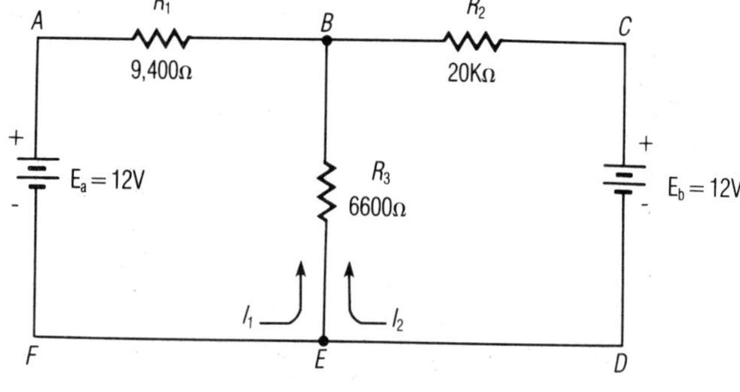

Fig. 3-27

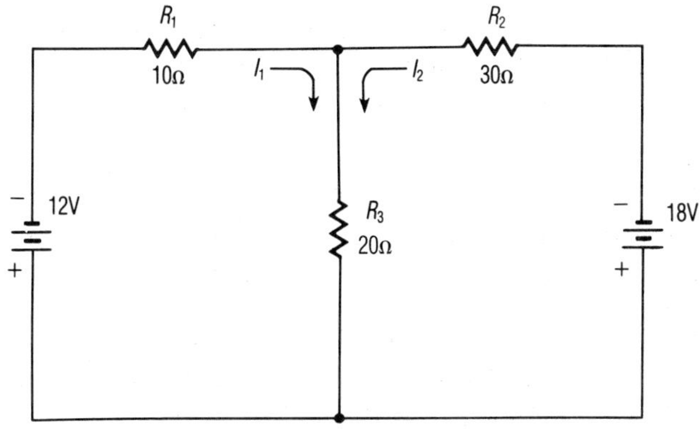

Fig. 3-28

1. Kirchhoff's first law states that the current entering a junction is equal to the current leaving the junction.
2. Establish current loops. By inspection you find that I_1 is from the 12-volt battery's negative terminal to the positive terminal. This means I_1 flows upward through R_3 and then through R_1 back to the positive terminal of the battery.
3. Establish that I_2 flows from the 18-volt battery's negative terminal up through R_3 and then through R_2 and back to the positive terminal of the battery.
4. Keep in mind that the 12-volt flow from the battery on the left (E_a) is distributed across R_3 and R_1. Then set up your equations:

Equation 1	**Equation 2**
a. $12 = 10(I_1) + 20(I_1 + I_2)$	$18 = 30(I_2) + 20(I_1 + I_2)$
b. $12 = 10I_1 + 20I_1 + 20I_2$	$18 = 30I_2 + 20I_1 + 20I_2$
c. $12 = 30I_1 + 20I_2$	$18 = 20I_1 + 50I_2$

5. Now that you have the two equations reduced to their lowest possible terms, it is time to eliminate one of the unknown currents—either I_1 or I_2.
6. If you decide to eliminate I_1, place one equation below the other:

$$12 = 30I_1 + 20I_2$$
$$18 = 20I_1 + 50I_2$$

By inspection you can see that I_1 and I_2 cannot be eliminated by mere subtraction. It is necessary to multiply one of the equations by a given number selected to produce an equal number below and above. In order to do this, it is necessary to multiply the 9-volt equation by -1.5, which will produce the following:

$$\begin{aligned} 12 &= \cancel{30I_1} + 20I_2 \\ -27 &= -\cancel{30I_1} - 75I_2 \\ \hline -15 &= \qquad\quad -55I_2 \end{aligned}$$

$$I_2 = \frac{15}{55}$$

$$I_2 = 0.2727272727 \text{ amperes}$$

7. Substitute the I_2 value in either equation to produce I_1 for the circuit.

$$12 = 30I_1 + 20I_2$$
$$12 = 30I_1 + 20(0.2727272727)$$
$$12 = 30I_1 + 5.4545454545$$
$$30I_1 = 6.5454545454$$
$$I_1 = \frac{6.5454545454}{30}$$
$$I_1 = 0.21818181818 \text{ amperes}$$

8. The current through R_3 is the sum of the two currents, or:

$$\begin{aligned} I_1 &= 0.21818181818 \\ +I_2 &= 0.27272727272 \\ \hline I_1 + I_2 &= 0.49090909090 \end{aligned}$$

9. Since the answers come out positive, the assumed direction of upward through R_3 is correct for current flow.
10. Check to see if your equations and your work are correct by doing the following:

    a. Take the equation and state it as you first arrived at it:

$$12 = 10(I_1) + 20(I_1 + I_2)$$

b. Substitute the values of the currents where called for:

$$12 = 10(0.218181818) + 20(0.4909090909)$$
$$12 = 2.18181818182 + 9.81818181818$$
$$12 = 12$$

11. To make sure everything checks out correctly, just substitute for the other equation to see if you get the proper voltage drop across R_2. Then check to see if $E_{R_2} + E_{R_3}$ equals 18 volts. If so, you know that your work on the problem is correct.
12. Check your answers. Another quick check shows that:
    a. R_3 has a voltage drop of 9.818181818 volts.
    b. R_2 has a voltage drop of 8.181818181 volts.
    c. R_1 has a voltage drop of 2.181818181 volts.

PROBLEMS

In Figure 3-29 the voltages differ and resistors differ in value as indicated in the charts below. Fill in the missing values for each of the problems.

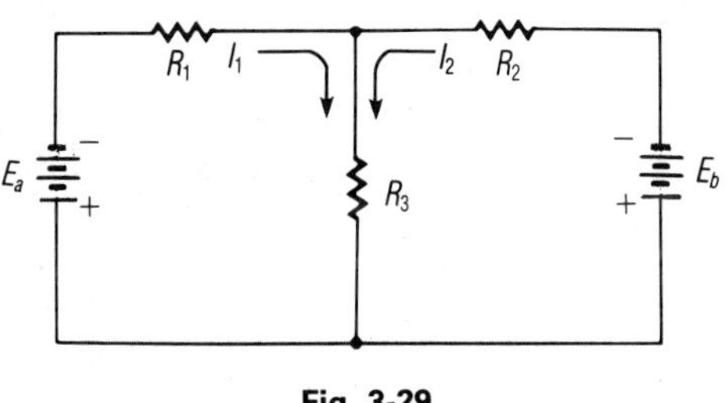

Fig. 3-29

1.

R_1	R_2	R_3	I_1	I_2	E_{R1}	E_{R2}	E_{R3}	E_a	E_b
10 Ω	6 Ω	7 Ω	?	?	?	?	?	11V	8.1V

2.

R_1	R_2	R_3	I_1	I_2	E_{R1}	E_{R2}	E_{R3}	E_a	E_b
4,000Ω	1,200Ω	1,400Ω	?	?	?	?	?	219v	161v

CHAPTER 4

Meters

THE AMMETER

The D'Arsonval meter movement is one that has a moving coil. The permanent magnet is stationary. Current flows through the coil and sets up a magnetic field that is as strong as the amount of current and number of turns in the coil make it. Since the number of turns is fixed, the only variable in the meter is the amount of current flowing through the coil. The strength of the magnetic field, then, is directly related to the amount of current that flows through the coil. This magnetic field is brought into the permanent-magnet field created by the permanent magnet of the meter. See Fig. 4-1.

In order for the meter to be very accurate, much attention to detail is necessary. The coil of wire has to be balanced and has to be able to swing with the strength of the magnetic field created by the current through it.

It is very important to fit the meter to the job. For example, a 1-milliampere meter is ideal for measuring direct currents up to 1 milliampere (0.001 A). But such a meter does not have sufficient sensitivity to measure currents of less than 100 microamperes (0.1 mA).

Meters are most accurate in the middle 80% of their range. Their accuracy falls off in the lowest and highest 10% of the scale.

Shunts

The 1-milliampere (1 mA) meter can be used to measure direct current that is larger than its highest range. To do this, however, a *shunt* is needed to bypass some of the current, or all that is over

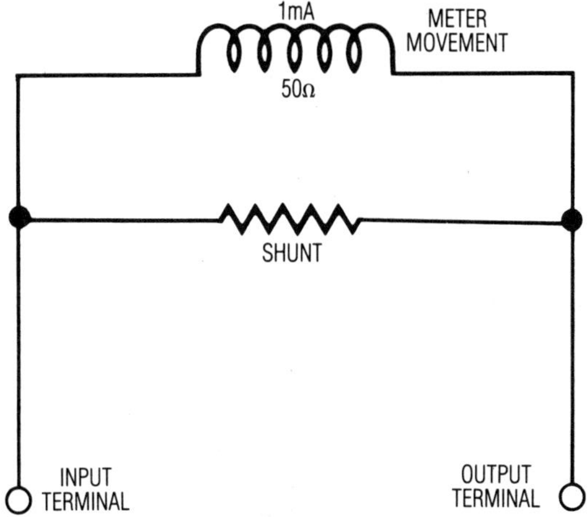

Fig. 4-1

full scale for the meter. If, for instance, you wanted to measure 2 mA with a 1-mA meter movement, you would need some way of protecting the meter movement so it is not burned out. This can be done by adding a resistor (shunt) in parallel with the meter movement. Fig. 4-2 shows a shunt located across a meter movement.

Check out Fig. 4-2 and see that the meter movement is capable of handling 1 mA or (0.001 A). If you want to make it handle twice as much, you have to make sure the shunt is the proper size to take the other milliampere of current.

Figuring out shunts is easy since the circuit is nothing more than a parallel configuration. That means the voltage is the same across the meter movement and the shunt. It also means that the current divides according to the resistance. If the meter movement has 50 ohms internal resistance and will allow 1 mA of current through it, you can see what the voltage drop is by using Ohm's law.

$$E = I \times R$$
$$E = 0.001 \times 50$$
$$E = 0.05 \text{ volts}$$

Next, you want to know the size of the shunt. You can do that because the current through the shunt is everything that will not

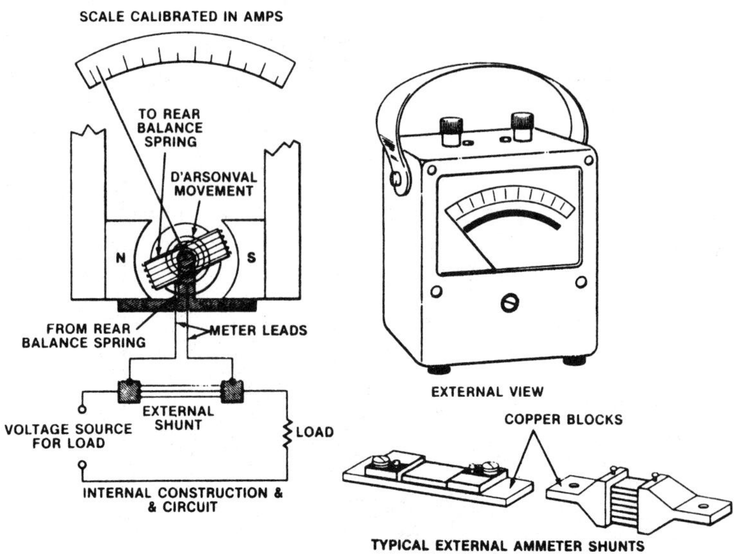

Fig. 4-2

go through the meter movement. If the meter movement is limited to 1 mA and you have 5 mA in the circuit you are measuring, you know that the shunt has to handle the extra 4 mA to protect the meter movement from an overload or too much current.

Now you know that the current through the shunt is 4 mA (or 0.004 A). You also know the voltage drop across the shunt is 0.05 volts, since that was the voltage you figured out for the meter movement and the two are in parallel. Resistors in parallel have the same voltage drop across them. Now you have what you need to find the resistance of shunt. Simply use Ohm's law:

$$R = \frac{E}{I}$$

$$R = \frac{0.05}{0.004}$$

$$R = 12.5 \text{ ohms}$$

Now let's try another problem and see if we can come up with the size of shunt needed.

Example 1

A 1-mA meter movement with 100 ohms internal resistance is used to measure 1 ampere in a circuit. What would be the size of shunt needed to prevent the meter movement from being damaged?

1. You know the current in the circuit to be 1 ampere. You know the meter can only take 1 mA. (The ampere is equal to 1,000 mA.)
2. First thing to do is find the voltage drop across the meter movement since you already know the resistance and current through it.

$$E = I \times R$$
$$E = 0.001 \times 100$$
$$\mathrm{E} = 0.1 \text{ volt}$$

3. Now what is the shunt size? You know the voltage drop across the shunt is the same as across the meter movement, or 0.1 volt. You know also that 1 mA goes through the meter movement and the rest through the shunt. That means that 999 mA go through the shunt. This 999 mA is the same as 0.999 A. We convert to amperes since the voltage is given in volts.
4. The shunt is equal to:

$$R = \frac{E}{I}$$
$$R = \frac{0.1}{0.999}$$
$$R = 0.1001 \text{ ohm}$$

As you can see, the size of the shunt is very small. In fact, it is a short circuit for all practical purposes.

PROBLEMS

Find the missing value in the following meter movements:

	Voltage Across Meter Movement	Current in Circuit	Current Through Meter Movement	Size of Shunt	Meter Movement Resistance (in ohms)
1.	?	1A	10mA	?	100Ω
2.	?	1A	1mA	?	1000Ω
3.	?	10A	1mA	?	50Ω
4.	?	1A	100mA	?	5000Ω

VOLTMETER

The voltmeter is made from an ammeter or milliammeter movement. It is connected somewhat differently than the ammeter so it can take the higher voltages. To limit the amount of voltage that gets to the meter movement, the *multiplier*, or limiting resistor, is placed in *series* with the meter movement. See Fig. 4-3.

If the same milliammeter movement is used here as in the milliammeter, we can say the meter movement has 1 mA of current through it. That means no more than 1 mA can flow since it would damage the movement if more were allowed in the circuit. The current limitations of the coil in the meter movement, then, are the limiting factors here. We use the 1 mA throughout our calculations since this is a series circuit and the multiplier is placed in series with the movement to to protect it. If the meter movement has 100 ohms resistance, it means the meter movement will have 0.1 volt across it when there is full-scale deflection. Keep in mind

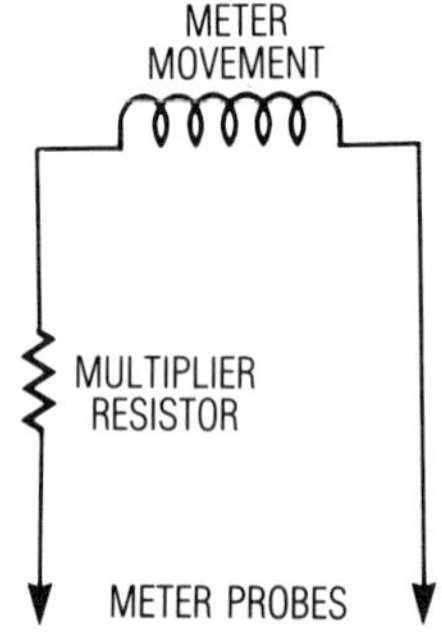

Fig. 4-3

the limitations of this meter movement when designing the meter as a voltmeter and using it to measure more than its 0.1-volt limit.

Suppose you want to measure 100 volts with this meter. The meter movement itself can take only 0.1 volt. That leaves you with 99.9 volts to drop across the multiplier in series with the movement. You already know that the current in the circuit is 0.001 A because that is the limit on the meter movement. Therefore, you can put Ohm's law to work to find the size of the multiplier needed to allow you to measure 100 volts.

$$R = \frac{E}{I}$$

$$R = \frac{99.9}{0.001}$$

$$R = 99{,}900 \text{ ohms}$$

That means the multiplier must be 99,900 ohms in order to measure the 100 volts.

Example 2

Suppose you want to measure 500 volts with the meter movement just described (1 mA with 100 ohms internal resistance).

1. You know that the current in the circuit is limited to that which the meter movement can handle. That is 0.001 A in this case.
2. You also know that you want to measure 500 volts.
3. $$R = \frac{E}{I}$$

   $$R = \frac{500}{0.001}$$

   $$R = 500{,}000 \text{ ohms}$$
4. You now know that the *total* resistance needed to limit 500 volts to 1 milliampere in a circuit is 500,000 ohms.
5. You know that the voltmeter circuit has two resistors in series. You also know that the movement resistance is 100 ohms. Since there are only two resistors in the circuit, just

subtract the 100 ohms from the total resistance of 500,000 ohms.

6. 500,000 − 100 = 499,900 ohms for the multiplier.

PROBLEMS

Fill in the blank spaces with the correct value.

	Movement Resistance	Movement Current	Voltage to Be Measured	Resistance of the Multiplier
1.	50Ω	1mA	100V	?
2.	500Ω	1mA	500V	?
3.	1000Ω	10mA	1000V	?
4.	500Ω	50μA	100V	?

SERIES OHMMETER

The ohmmeter is an instrument that indicates the resistance value of a circuit element or network on a scale calibrated in ohms. It is used also to locate shorted or open circuits, check circuit continuity, and provide a rough check on capacitors. The ohmmeter is one of the basic test instruments, along with the voltmeter and ammeter.

The ohmmeter consists of a sensitive current meter, a source of low-voltage d.c., and some form of current-limiting resistor. The meter usually is a conventional direct-current moving coil type, and a battery supplies the necessary d.c. voltage. The values of resistance encountered in electronic equipment vary from fractions of an ohm to many millions of ohms (megohms). Most ohmmeters are constructed to cover a wide range, from very low to very high resistance, by connecting various values of current-limiting resistors in the ohmmeter circuit. The different ranges can be selected as required by means of individual input terminals or a selector switch. When a selector switch is used, only one set of input terminals, common to all ranges, is necessary. The meter scale can be calibrated to read each individual range directly, or one meter scale can be used and a multiplying factor applied for each range.

Series Ohmmeter

In the basic ohmmeter circuit of Fig. 4-4, a 4.5-volt battery, a variable resistor, R_A, and a fixed resistor, R_B, are connected in series with a milliammeter. The two leads with test prods are connected across the resistance to be measured, R_x. The meter is a 0-to-1-mA movement, requiring 1 mA of current for full-scale deflection. Internal resistance of the meter movement, R_m, is 50 ohms. The fixed resistor, R_B, limits the flow of current and is placed in the circuit to prevent damage to the meter. If no limiting resistor is placed in the circuit and the variable resistor is adjusted to a low value of ohms, the current flow in the circuit becomes excessive. The variable resistor, R_A, adjusts the series resistance in the circuit so that 1 mA flow when the test prods are shorted together. This occurs when the total series resistance is 4,500 ohms. Ohm's law says:

$$R = \frac{E}{I}$$

$$R = \frac{4.5}{0.001}$$

$$R = 4{,}500 \text{ ohms}$$

In the circuit in Fig. 4-4 the internal resistance of the meter is 50 ohms, the resistance of R_B is 4,000 ohms, and R_A is adjusted to 450 ohms, making up the required series resistance of 4,500 ohms. This gives full-scale deflection on the 0-to-1-mA meter.

When the test prods are shorted, current flows through the circuit and the meter needle is deflected across the scale. A knob on the front of the meter panel is called the *Ohms Zero Adjust* control. This is a resistor (R_A) that is adjusted to provide full-scale deflection of the meter needle. This position of the needle corresponds to zero resistance, since the two connected prods are short-circuited across the ohmmeter terminals. In the series-type ohmmeter, full-scale deflection of the meter pointer indicates the lowest resistance. The opposite end of the scale represents the highest resistance.

When the battery ages, the available voltage decreases and the variable resistor compensates for this condition. The current can be

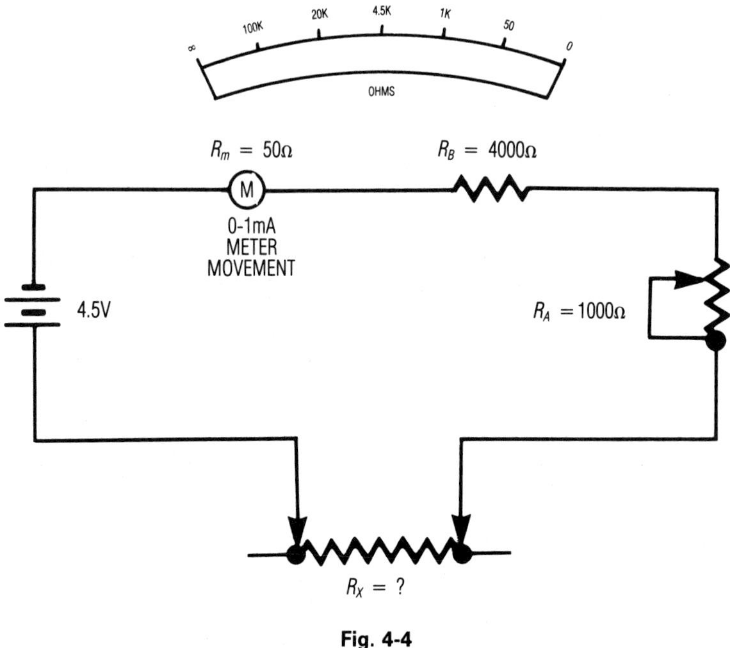

Fig. 4-4

increased to the required value by adjusting R_A, which decreases the total series resistance. The zero-adjusting resistor, R_A, can be connected either in series or in shunt with the meter. When it is connected in shunt, a fixed resistor is connected in series with R_A, and both resistors are connected across the meter. For accuracy in meter readings, a zero-ohm adjustment is made each time the ohmmeter is used.

After the ohmmeter has been adjusted for full-scale deflection, the test prods are separated and the meter pointer returns to the open circuit position on the left of the scale. Placing the prods across the unknown resistor connects it in series with the ohmmeter circuit, the current is reduced proportionately, and the meter pointer no longer deflects full-scale. If the value of R_x is equal to the combined resistance of the current-limiting resistor, R_A, the internal resistance of the meter, the total circuit resistance then becomes 4,500 plus 4,450 plus 50, or 9,000 ohms. The current in the circuit is now

$I=\frac{E}{R}$, or $\frac{4.5}{9{,}000}=0.0005$ A, which is the same as one-half of a milliampere (0.0005). This is half of the 1 mA required for full-scale deflection, and the meter pointer is deflected only half-scale.

When calibrating the meter scale of the ohmmeter in Fig. 4-4, the half-scale deflection point is marked 4,500 ohms. If the value of the unknown resistor, R_x, is twice the resistance of the ohmmeter, the total circuit resistance is tripled, and the current is reduced to one-third the full-scale deflection value. The meter pointer deflects to one-third full scale and corresponds to an R_x of 9,000 ohms.

These and other points on the ohmmeter scale can be calibrated conveniently from the formula:

$$R_x = R_c \frac{(I_1 - I_2)}{I_2}$$

R_x is the unknown resistance

R_c is the total circuit resistance when the prods are shorted together

I_1 is the ohmmeter current with prods shorted

I_2 is the ohmmeter current with the prods across R_x

Example 3

Find the value of R_x when the pointer deflection indicates a flow of 0.25 mA through the ohmmeter circuit. The value of R_x is now:

$$R_x = 4500 \frac{(1 - 0.25)}{0.25} = 13{,}500 \text{ ohms}$$

By use of this formula the entire meter scale can be calibrated to read directly in ohms the value of any unknown resistor.

PROBLEMS

Using Fig. 4-4 as the meter movement and its necessary components, find R_x and fill in the chart below, given the following values:

	R_x	R_c	I_1	I_2
1.	?	4,500	1 mA	0.3 mA
2.	?	4,500	0.5 mA	0.25 mA
3.	?	4,500	0.5 mA	0.125 mA
4.	?	4,500	0.5 mA	0.0625 mA
5.	?	4,500	0.75 mA	0.25 mA

SHUNT OHMMETER

The shunt ohmmeter is used to measure low resistances such as those encountered in home appliances. The resistances of a toaster and electric motor are very low. The shunt ohmmeter is a compact device used to test devices with low resistance. It is portable because it has its own power source.

The basic concept of the shunt ohmmeter is that when two resistances of equal value are connected in parallel, the current divides between the two resistances, with half flowing in each. That means if a 50-ohm resistor is connected in parallel with a meter movement whose internal resistance is also 50 ohms, the current flowing through the circuit will divide between the two resistances, with the meter having only half of the total current flowing through it. If the circuit supplying the current is adjusted for a current of 1 mA and the 50-ohm resistor is connected across the meter movement—see Fig. 4-5—the current through the meter would drop to 0.5 mA. The other 0.5 mA will flow through the added resistance.

Since the shunt ohmmeter is *on* all the time, as far as the circuit is concerned, it must have an on-off switch so the battery is not exhausted when the meter is stored.

In Fig. 4-6 the total resistance of the circuit without the unknown resistor, R_x, is 3,000 ohms. If there is a 3-volt battery, it means a current of 1 mA will flow. When the unknown resistor is placed across the meter movement, it becomes a shunt for the meter

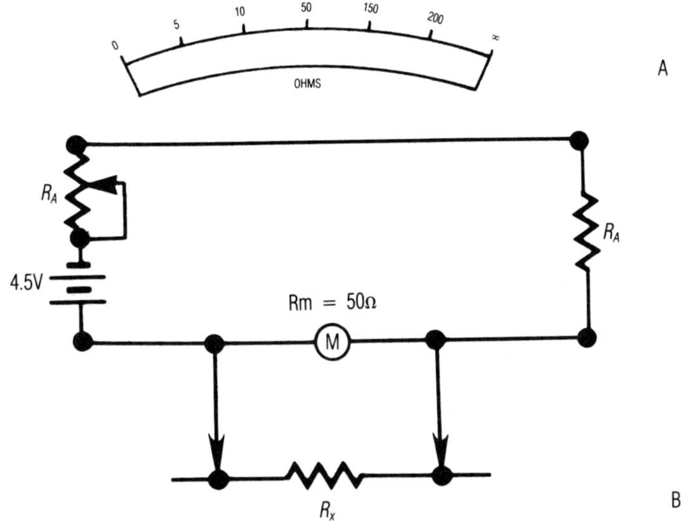
0
5
10
50
150
200
∞
OHMS
A
R_A
R_A
4.5V
Rm = 50Ω
M
R_x
B

Fig. 4-5

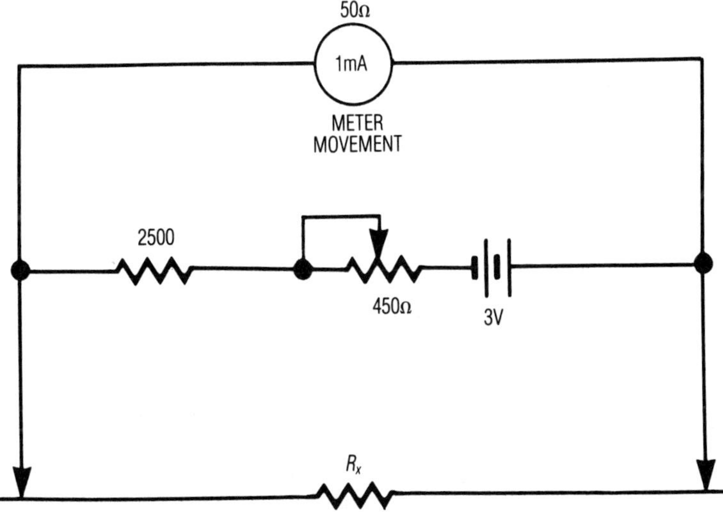
50Ω
1mA
METER
MOVEMENT
2500
450Ω
3V
R_x

Fig. 4-6

movement, That means the current will divide between the two resistances. If the unknown resistance is the same size as the meter movement resistance, the current flow through the meter will be only 0.5 mA. If you are calibrating the meter to read resistance, you would then mark 50 ohms for the middle of the meter scale.

The table in Fig. 4-7 shows how the current divides in the shunt part of the meter. It is a ratio. We know that the meter is made up of a series-parallel circuit. However, if you adjust the

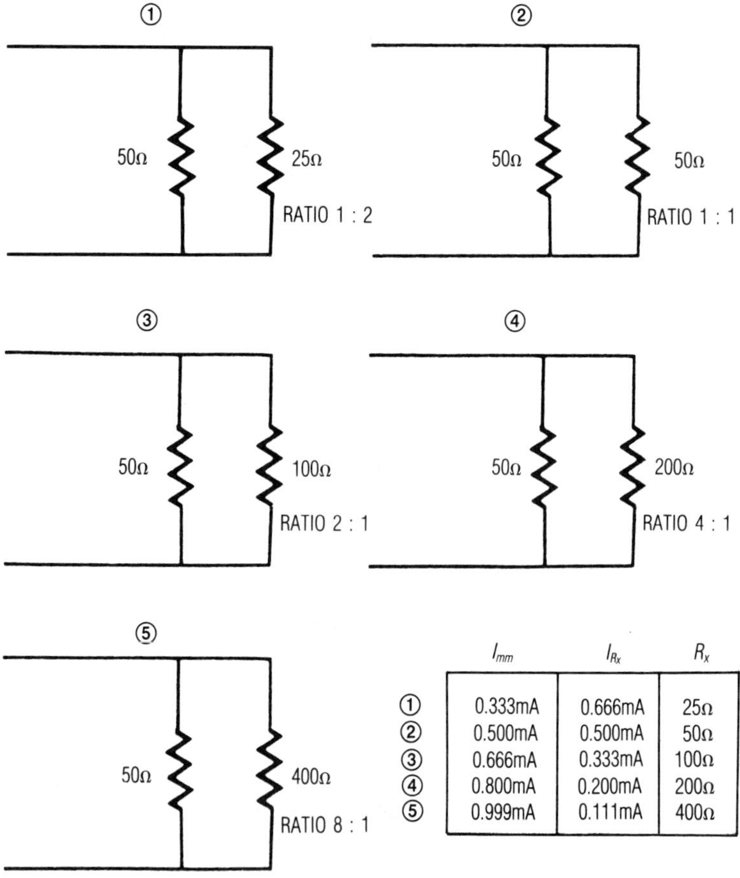

	I_{mm}	I_{Rx}	R_x
①	0.333mA	0.666mA	25Ω
②	0.500mA	0.500mA	50Ω
③	0.666mA	0.333mA	100Ω
④	0.800mA	0.200mA	200Ω
⑤	0.999mA	0.111mA	400Ω

Fig. 4-7

circuit so that the current through the meter is 1 mA, it will increase slightly as the 50-ohm movement is shunted by an unknown. However, this increase is in the microamperes and can't be detected too well on a 1-mA meter movement. Therefore, if we assume that the unknown resistor is placed, as in Figure 4-6, directly across the meter movement, it is a simple two-resistor parallel arrangement. The ratio determines the amount of current that goes through each resistor. The resistor with the smaller resistance value will have more current (due to Ohm's law).

Now take a look at Fig. 4-7 and the five examples of resistors across the 50-ohm resistor. The first one is a ratio whereby the resistance (R_x) is less than the meter movement resistance. That means twice as much current will flow through the 25-ohm resistor than through the 50-ohm resistor. However, in the second part of Fig. 4-7 you see that the resistances are the same. That means the current through the meter will drop to 0.5 mA since half will go through the unknown resistance. In #3 the resistance is larger than the meter movement resistance. That means that twice as much current goes through the meter movement than through the unknown resistor. That also means that the needle on the meter will swing more toward the right, to indicate a larger resistance.

As you can see from these examples, the shunt ohmmeter is limited in its ability to measure resistances. It is very good for 10 to 200 ohms.

PROBLEMS

Suppose we change the resistances and the voltage of the battery but keep the same meter movement. Can you figure out the current through the meter movement (I_{mm}) and the current through the unknown resistance (I_{R_x})? See Fig. 4-8 for the circuit.

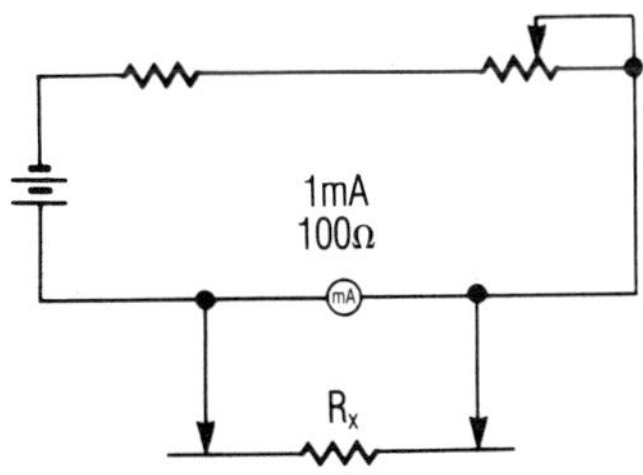

Fig. 4-8

	I_{mm}	I_{R_x}	R_x
1.	? mA	? mA	50
2.	? mA	? mA	100
3.	? mA	? mA	200
4.	? mA	? mA	300
5.	? mA	? mA	400

CHAPTER 5

Alternating Current and Inductance

ALTERNATING CURRENT

Alternating current (a.c.) is an electric current that moves first in one direction for a fixed period of time and then in the opposite direction for the same period of time. Alternating current builds up to a maximum in a positive direction, and then falls off to zero value again (the condition of no current flow) before building up to a maximum in the opposite, or negative, direction and then falls again to zero. See Fig. 5-1.

Electromagnetic Induction

Direct current (d.c.) has been around much longer than alternating current, having been the first type of electric current generally used. It was created by using wet cells and batteries. In 1819 Hans Christian Oersted, a Danish physicist, was experimenting with d.c. and accidentally discovered that a wire carrying an electric current affected a compass needle and was, therefore, itself a kind of magnet. This was called an electromagnet to distinguish it from a natural or artificial magnet. In either case, the magnetic lines of force, or the force field, about a wire or coil carrying a current is the same as that about a natural magnet. Oersted's discovery meant that electricity and magnetism were closely related, since one could be used to produce the other. However, it was not until 1831, 12 years later, that Michael Faraday in England and Joseph Henry in the United States were able to prove that a magnet could be made to produce an electric current.

Faraday, in his now classic experiment, connected a sensitive

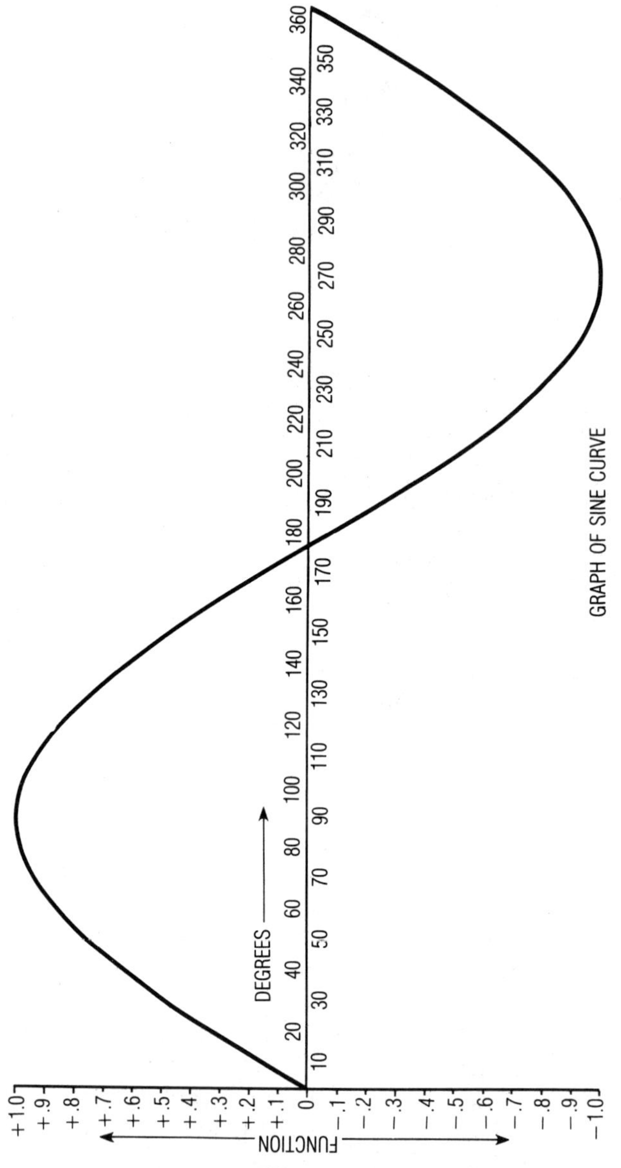
DEGREES
0 10 20 30 40 50 60 70 80 90 100 110 120 130 140 150 160 170 180 190 200 210 220 230 240 250 260 270 280 290 300 310 320 330 340 350 360
FUNCTION
+1.0 +.9 +.8 +.7 +.6 +.5 +.4 +.3 +.2 +.1 0 −.1 −.2 −.3 −.4 −.5 −.6 −.7 −.8 −.9 −1.0
GRAPH OF SINE CURVE

Fig. 5-1

galvanometer across a coil and found that when a magnet was thrust into the coil, a current flowed in the coil. He also found that when the magnet was withdrawn, a current flowed in the opposite direction. Current flow, however, resulted only during the time the magnet was *moving*—that is, when the lines of force or force field about the magnet cut the wires of the coil. The opposite condition was also found to be true—that is, if the magnet was held stationary and the coil moved, current flowed during the time of movement. Thus, an alternating current was produced. Figure 5-2 shows this principle in terms of a single conductor and a horseshoe magnet.

Producing Alternating Current

It is mechanically awkward to produce an alternating emf by moving a magnet first into a coil and then out again. Moving the coil back and forth about a stationary magnet is equally awkward. But a simple a.c. generator may be set up by rotating a single-turn coil within a stationary field. Thus, continuous uniform movement is possible, and the direction of the induced emf reverses as the coil turns, since as the conductor moves down through the field a voltage is induced in one direction, and induced in the other direction as it moves up through the field. Take a close look at Fig. 5-3 to see how a point with uniform circular motion produces a sine wave.

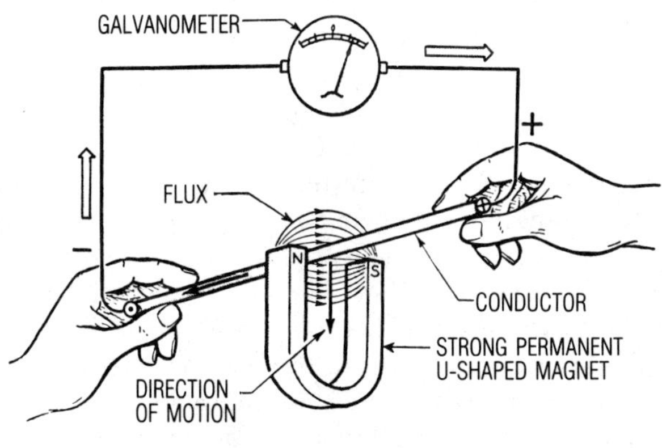

INDUCING AN EMF.

Fig. 5-2

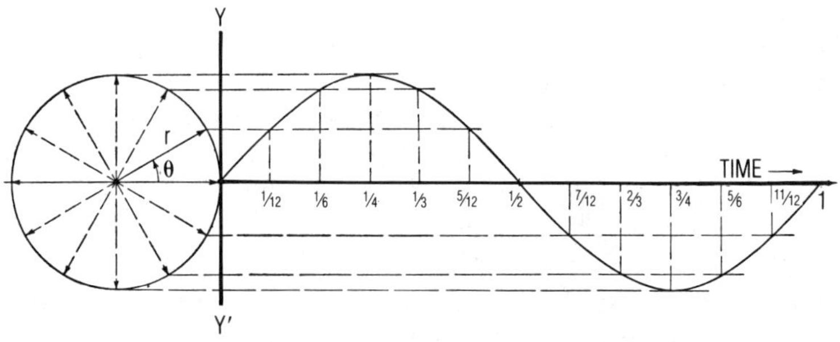

PROJECTION OF POINT HAVING UNIFORM CIRCULAR MOTION.

Fig. 5-3

This is the product of an alternator or alternating-current generator. The rotated point is the end of the conductor in a magnetic field as afforded by a generator.

Phase Difference: Leading and Lagging

The time it takes to produce one revolution of the a.c. generator is called its *cycle*. However, the frequency at which this generator is rotated is measured no longer in cycles per second (cps), but in *hertz* (Hz).

One cycle completed in one second equals one hertz. When the cycle has no time base, it is referred to as a cycle, as you would refer to a washing machine completing its cycle for washing clothes. Once the cycle is repeated more than once a second, it is measured in hertz, the measurement of frequency established some time ago and agreed upon internationally. Try to keep in mind that a generator's cycle is merely that: the time it takes to complete a 360° rotation. It is used here merely in the analysis of the sine wave and its components so it can be compared to direct current.

As you already know, Ohm's law was in use long before alternating current was established as a practical power source. Therefore, all the formulas and most of the work in the electrical field had to be revised when alternating current came along. That is why some of the values are found in the section on rms, peak, and average alternating-current values. They are simply made equal to d.c. so the mathematical work can be used for both ac and dc.

The time of a cycle, or one hertz of alternating current or voltage, is usually expressed in electrical degrees. See Figs. 5-1 and 5-3. That means 90° is said to be one-fourth of a cycle and represents an amount of time dependent on the frequency of the voltage, or the hertz.

If the voltage considered is the usual 60-Hz a.c. used in the home, 1 Hz is generated in 1⁄60 of a second. Then 90°, or 1⁄4 (0.25) of a hertz, actually represents 1⁄4 of 1⁄60 (0.01666666666) of a second —that is, 1⁄4 times 1⁄60 or 4.16666 milliseconds ($0.25 \times 0.01666666 = 0.0041666666$).

The *phase* of this voltage is also 90°, or 1⁄240 of a second, which is also referred to as 4.16 milliseconds. Phase then is defined as the difference in time between any point on a cycle and the beginning of that cycle. The beginning of a cycle generally is taken to be the point at which the cycle passes through zero moving in a positive direction. Such a consideration of phase, although seldom used in reference to a single voltage, is of immediate practical importance when two a.c. voltages or currents are present in the same circuit. It is necessary then to determine the position of one with relation to the other at any given instant in time.

Lead and Lag

If two 60-Hz a.c. generators are put in operation at the same time and connected to the same circuit, the two varying voltages rise and fall and reverse direction at the same time. These voltages are said to be *in phase*. See Fig. 5-4. If one generator is started 1⁄240 second, or 4 milliseconds, after the other, the two voltages do not rise and

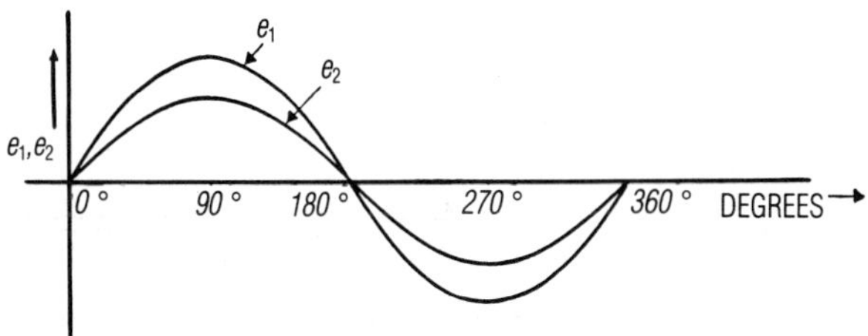

TWO SINE-WAVE VOLTAGES IN PHASE.

Fig. 5-4

fall together but are separated by a definite period of time that may be expressed in degrees. The voltages from the two generators are then said to be *out of phase*. The first is said to *lead* the second, or the second is said to *lag* the first, by the number of degrees expressing this time difference, that is, by 90°.

In Fig. 5-5 this condition is shown as voltage e_2, which lags voltage e_1. This means that the starting point of e_2 is 90° to the right of e_1. Over the time interval 180° to 270°, e_2 is positive and e_1 is negative. Note that the x-axis, or time axis, moves from left to right, and therefore any point to the *right* of any other point is later in time, or *lags*, the other.

Figure 5-6 shows two voltages 180° out of phase. Both voltages go through their zero points and maximum points at the same time, but e_1 is in the opposite direction from e_2—that is, they are always of the opposite sign. One is negative when the other is positive.

If these two voltages are of equal value, or amplitude, and present in the same circuit, the resultant voltage is zero, since the voltages cancel completely. However, if the two voltages are in

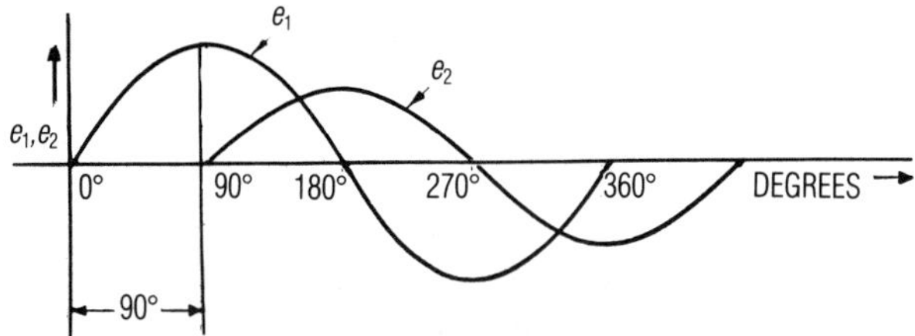

TWO SINE-WAVE VOLTAGES 90° OUT OF PHASE.

Fig. 5-5

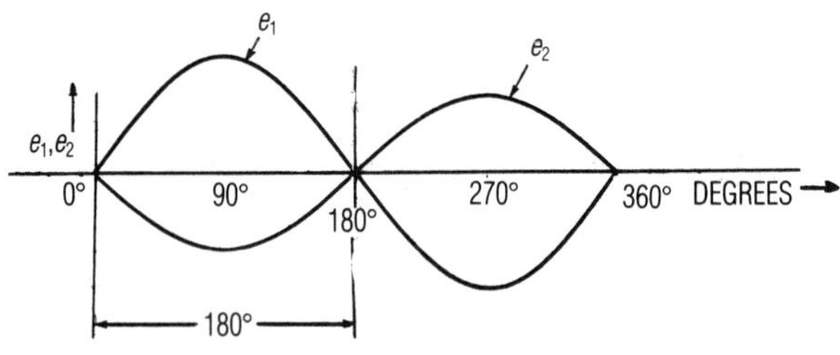

TWO SINE-WAVE VOLTAGES, 180° OUT OF PHASE.

Fig. 5-6

phase, as shown in Fig. 5-4, the resultant voltage is the sum of the two voltages. For out-of-phase relationships less than 180°, the resultant voltage is the vector sum of the two voltages.

Phase Difference

Please note that when the time difference in starting the two previously mentioned 60-Hz generators is 1/60 of a second, or 1 Hz, the voltages remain in phase. But a time difference of 1.25 Hz, for instance, is expressed as the fractional part of 1 Hz, or a 90° phase difference. From this it follows that if the time difference is any whole-number multiple of 1 Hz, the voltages are in phase, and that in any other case, the time difference is expressed simply as the fractional part of a single hertz. In addition, a phase difference is usually expressed in degrees from 0° to 180° because any angle greater than 180°—for instance, a 210° *lead* by the first generator—may be expressed in terms of the second generator as a 150° *lag*. As you can see, lead and lag are in reference to something. If the reference is one generator, it may be lead and if another, it may be lag. The reference generator must be identified in order to make the lead and lag known.

Example 1

These points may now be expressed generally and illustrated by Fig. 5-7. Two a.c. generators are put in operation at different times. Generator 2 is started t seconds later than generator 1. Both machines deliver sine-wave voltages at the same frequency. In Fig. 5-7, A is a graph of the voltage of generator 1, B is a graph of the voltage of generator 2, and C is a graph of the voltages of both generators plotted on the same time base. Note that voltage e_2 does not begin until a definite time interval (t_0 seconds) has elapsed since the beginning of voltage e_1. Thus, if t_0 equals 1 millisecond, then voltage e_2 will attain the same value after 2+1 or 3 milliseconds. Figure 5-7, then, shows the following important points:

1. The zero points of e_1 and e_2 are separated by t_0 seconds and the maximum points are separated by t_0 seconds.
2. Voltage e_1 goes through its zero value t_0 seconds before t_2.
3. The graph of e_2 can be reproduced by shifting that of e_1 to the right t_0 seconds.

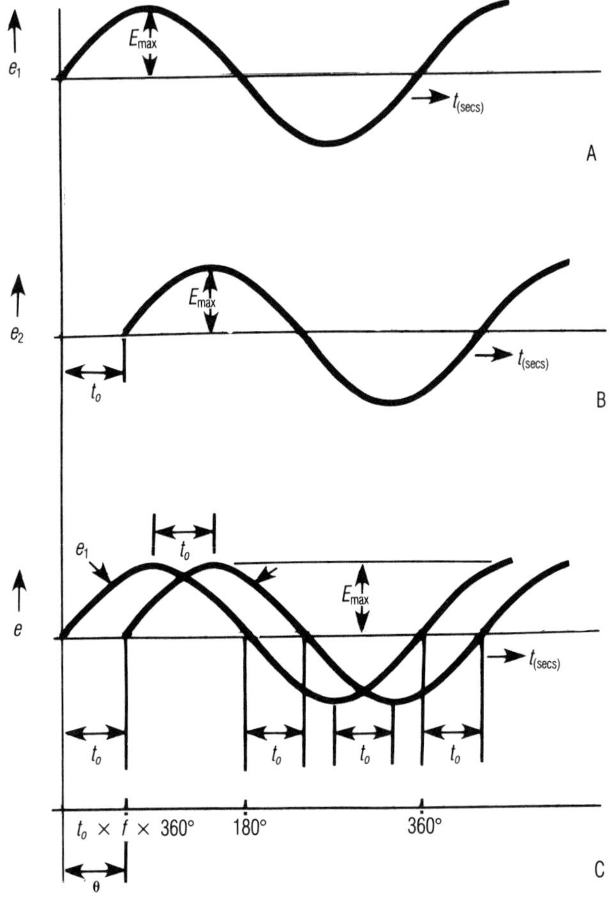

PHASING OF A.C. GENERATORS.

Fig. 5-7

The time lag of e_2 (t_0 seconds) may be converted to a lag in electrical degrees by the following equation:

$$\text{Electrical degrees} = t_0 \times \text{frequency} \times 360°$$

If two 60-Hz a.c. generators are started 1 millisecond (0.001) apart, the phase shift in seconds is equal to 10^{-3}. Then the phase shift in electrical degrees is equal to $10^{-3} \times 60 \times 360°$. This equals 21.6°. The second generator to start is then said to lag the first by 21.6°. See Fig. 5-8.

Example 2

If three generators are started at different times, each generator leads or lags the other two generators by a specified time. Figure 5-9 shows this condition. Generator 1 is started first, generator 2 next, and generator 3 last. Then voltage e_2 lags e_1; e_3 lags e_1 and also lags e_2. From the graph you get: $t_{3,2} = t_3 - t_2$.

For example, three 60-Hz a.c. generators are put into opera-

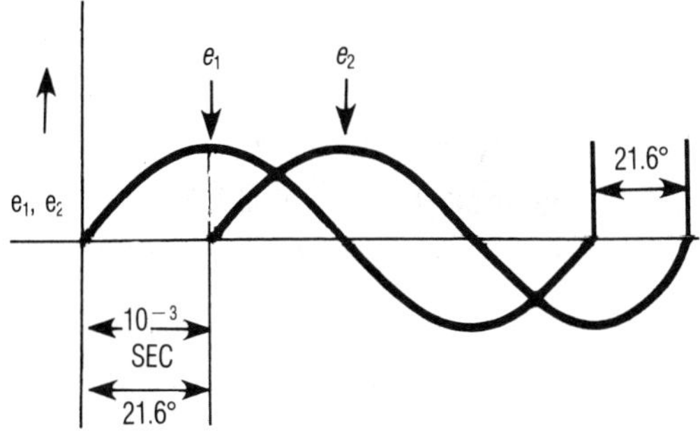

TWO SINE-WAVE VOLTAGES, 1-MILLISECOND DIFFERENCE. *10^{-3} IS SAME AS 0.001 SECOND.*

Fig. 5-8

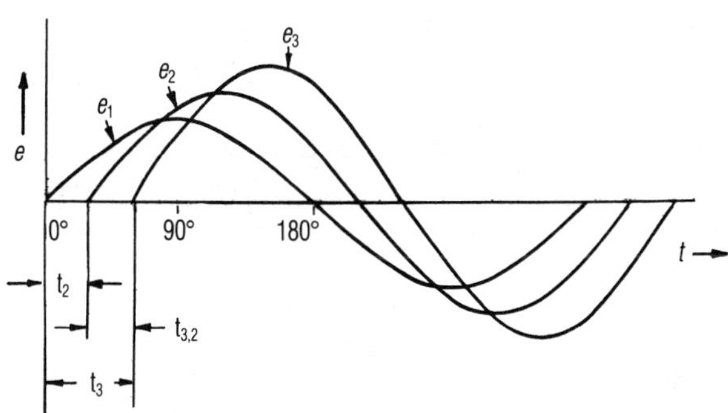

THREE SINE-WAVE VOLTAGES,PHASE DIFFERENCE

Fig. 5-9

tion. Generator 2 is started 2 milliseconds after generator 1, and generator 3 is started 1 millisecond after generator 2. Using the formula described in the preceding example, these time differences may be converted to equivalent electrical degrees of phase difference.

$$t_2 = 2 \text{ milliseconds, or } 0.002$$
$$t_{3,2} = 1 \text{ millisecond, or } 0.001$$
$$t_3 = t_2 + t_{3,2}\text{; then}$$
$$t_3 = 0.002 + 0.001 = 0.003 \text{ second}$$

Generator 2 lags generator 1 by $0.002 \times 60 \times 360°$.

So, when you multiply you get $0.002 \times 60 \times 360° = 43.2°$

Generator 3 lags generator 2 by $0.001 \times 60 \times 360°$

So, when you multiply you get $0.001 \times 60 \times 360° = 21.6°$

Generator 3 lags generator 1 by $0.003 \times 60 \times 360°$.

So, when you multiply you get $0.003 \times 60 \times 360° = 64.8°$.

PROBLEMS

Five generators are put on-line. Each is delayed 1 millisecond from the other in order. What is the phase difference (in degrees) as each is put on-line? Keep in mind that generator 2 is delayed 1 millisecond, generator 3 is delayed 2 milliseconds, generator 4 is delayed 3 milliseconds, etc.

	Generator	Amount of Delay from Generator 1
1.	2	
2.	3	
3.	4	
4.	5	

INDUCTANCE

The unit of measurement of inductance is the henry (H). It was named after Joseph Henry, the codiscoverer with Faraday of the principle of electromagnetic induction. A henry is defined as the inductance of a circuit in which a current *change* of 1 ampere per second causes a counter emf of 1 volt.

Inductance is that property of an electrical circuit that tends to prevent a change of current. When we looked at d.c. circuits, we found that they had a property called resistance. That resistance was the opposition to the flow of electrons, or current. The current in a typical d.c. circuit may be compared to an object in motion, such as an automobile, which is retarded only by the friction or resistance of the surface on which it moves. If, however, a d.c. circuit is broken suddenly, by opening a switch, for instance, a considerable spark jumps across the contacts of the switch as it opens. It may be said then that opening a d.c. circuit is like suddenly stopping an object in motion. But, from Newton's first law of motion it is known that an object in motion tends to remain in motion, and that a considerable force must be exerted to bring it to a stop. In the case of a speeding automobile stopped suddenly by a stone wall, this inertia, or the momentum that tends to keep the car moving, will smash the car and dissipate itself as heat. In the case of a d.c. circuit suddenly opened, particularly one carrying heavy current, the inertia, or momentum, of the current meeting the very high resistance of the open circuit produces a high voltage and dissipates itself as heat in a big spark.

Newton's first law of motion states that an object at rest tends to stay at rest unless acted on by an external force. For example, an automobile must exert considerable power in order to start. After reaching speed, the only power the car needs to keep it going is the power used to overcome friction. In like manner, an electric current cannot be started instantaneously; there is a delay between the application of the voltage and the rise of the current to its maximum amount. This effect of inertia in a d.c. circuit generally is not noticed, since the starting delay is slight; and, except in circuits carrying large amounts of current, the spark resulting from stopping the current, or opening the circuit, is not dangerous. It should be noted, however, that in d.c. circuits carrying heavy power, provision must be made to open the circuit by gradually increasing resistance until the current reaches a safe value.

In an a.c. circuit, however, this effect of electrical inertia, or inductance, is everpresent, since by definition an alternating current is one constantly changing in magnitude and periodically changing direction. Thus, if the current in a circuit is increasing, the inductance of the circuit is defined as that property of the circuit that tends to prevent the increase. If the current is decreasing, the inductance of the circuits tends to prevent the decrease. The greater the inductance of the circuit, the greater the opposition to a change in current.

Counter Electromotive Force (CEMF)

In a.c. circuits there is an everpresent opposition to the flow of current other than the d.c. resistance of the circuit. This additional opposition is caused by a *counter*, or back, emf. D.c. resistance is caused by the measurable opposition that can be read on an ohmmeter. This is called cemf to distinguish it from the applied voltage, which is the original force tending to set up a current flow. Counter emf is analogous to the force that opposes a change in motion of a mass. Stated generally, it may be said that the velocity of a mass cannot be changed instantaneously, whether that mass be an automobile, an electron, or an electrostatic charge.

In an electric circuit, the cemf is an induced emf, or voltage. This voltage is induced in conductors of the circuit not by means of an external magnetic field, as in the case of a simple two-pole generator, but by means of the magnetic field already surrounding any conductor carrying a current. Any change in current changes the intensity of this magnetic field, and the resultant emf induced, the counter emf, is a *self-induced* voltage. Thus, the property of a circuit that produces such an emf is called self-inductance. Actually, all elements in a circuit, including connecting wires, show some self-inductance, but for all practical purposes only those elements designed to make use of this property to advantage are known as inductances or inductors. Moreover, it may be said that counter emf is present in any a.c. circuit, but its effect is negligible in a circuit of moderate power, such as an electric lamp, which uses almost pure resistance as a load. But the effect of cemf is considerable in circuits (even of very low power) that use an inductance as part of the load, such as the primary of the power transformer in an ordinary television set.

Lenz's Law

H. F. E. Lenz deduced from the principle of the conservation of energy that the emf self-induced in a conductor carrying current is a counter emf. If the self-induced emf was not a counter emf, then an increase of current would aid the applied voltage, and this increase in applied voltage would in turn tend to increase the current. This process would continue, of course, until current reached an infinite amount, a condition not possible in the physical universe. Lenz's law states: *An induced emf always has a direction that opposes the action that produces it.*

Thus, when a current flowing through a circuit varies in magnitude, it produces a varying magnetic field that sets up an induced emf that opposes the current change producing it. Or, it may be said that when the current in a circuit is increasing, the induced emf opposes the applied voltage and tends to keep the current from increasing. When the current is decreasing, the induced emf aids the line voltage and tends to keep the current from decreasing.

This can be observed experimentally. Connect a power transformer to a 120-volt, 60-Hz line. If you check the resistance of the primary of the transformer before it is connected in the circuit, you will find it has 10 ohms resistance d.c. If you hook up this circuit and place an a.c. ammeter in the circuit you will find that it will not read

$$I=\frac{E}{R}$$

$$I=\frac{120}{10}=12 \text{ amperes}$$

as you would expect it to with its 10 ohms of resistance. Instead, the transformer will read about 1 ampere. There must then be something other than the resistance holding back the current. Hence, Ohm's law for d.c. circuits must be modified if it is to serve the needs here in an a.c. circuit with an inductor.

Keep in mind that the rate of change of the flux density is

equivalent to movement. But the flux density about a conductor is directly proportional to the current in the conductor. Therefore, the magnitude of self-induced emf depends directly on the rate of change of the current in the circuit. Thus, a rapidly changing current induces a greater counter emf than a slowly changing current. But for any a.c. the rate of change of current depends on the number of hertz, or the frequency. The counter emf then depends directly on the frequency.

The total magnitude of an induced emf depends also on the length of the conductor, since in the simple generator the length of the inductor and the number of conductors in a coil side determine the total emf induced. Thus, a long conductor has greater counter emf induced, or has more self-inductance, than a short one. If, however, a long conductor is wound on itself in the form of a coil, its self-inductance is increased because of the increase in total flux density. Such a coil, or inductance, is a solenoid, and the flux density about it may also be increased by the addition of a core material of high permeability, such as soft iron.

The magnitude of a counter emf, then, is determined by the following formula:

$$\text{cemf} = \frac{-0.4\pi N^2 \mu A}{l} \times \frac{\Delta i}{\Delta t}$$

N = number of turns
A = cross-sectional area in cm^2
μ = permeability of the core
l = length of coil in cm
L = self-inductance in henrys
Δi = change in current (amperes)
Δt = change in time (seconds)

Since the physical characteristics of the solenoid, called the geometery of the coil, are all contained in the factor $\frac{0.4\pi N^2 \mu A}{l}$, this expression may be isolated and given the name *inductance* and assigned the symbol L. That means that:

$$L = \frac{0.4\pi N^2 \mu A}{l}$$

Substituting L for this factor:

$$\text{cemf} = -L\frac{\Delta i}{\Delta t}$$

The minus sign means that the voltage developed is a counter voltage and opposes the force producing it. Also, it should be noted that the separation of the physical characteristics (inductance) from the rate of change of the current is done for convenience and is analogous to the separation usually made in the more familiar equation of the inertia of masses. That means the inductance can be defined as that property of a coil that opposes any change in circuit current.

Measurement of Inductance

Again, the unit of measurement of inductance is the henry (H). A henry is defined as the inductance of a circuit in which a current change of 1 ampere per second causes a counter emf of 1 volt. Since the henry is defined in terms of practical units, the factor 10^{-8} must be used if the cemf is to be read in volts and the rate of change of the current in amperes per second. That means the following modification of the formula has to be made:

$$L = \frac{0.4\pi N^2 \mu A}{l} \times 10^{-8}$$

This formula reveals the following important relationships:

1. The inductance of a coil is proportional to the *square* of the number of turns.
2. The inductance of a coil increases directly as the permeability of the material making up the core increases.
3. The inductance of a coil increases directly as the cross-sectional area of the core increases.
4. The inductance of a coil decreases as its length increases.

Measurement of CEMF

The formula derived previously for the magnitude of a cemf was found to be:

$$\text{cemf} = -L\frac{\Delta i}{\Delta t}$$

An examination of this formula reveals that the greater the inductance, or the faster the rate of change of the current, the greater is the cemf induced in the circuit.

Example 3

A coil of 1 H inductance has a current of 1 ampere flowing through it. If this current changes to 2 amperes in 1 second, what would be the cemf?

1. Find the formula with the proper symbols for what is given and what is needed to be found.
2. Substitute the values known into the formula:

$$\text{cemf} = -L\frac{\Delta i}{\Delta t}$$

$$\text{cemf} = -1 \times \frac{(2-1)}{1}$$

3. Solve to obtain: cemf $= -1$ volt.

Example 4

If the current change remains the same (2 amperes in 1 second) but the coil has an inductance of 10 H, what would be the cemf?

1. Find the formula with the proper symbols for what is given and what is needed to be found.
2. Substitute the values known into the formula:

$$\text{cemf} = -L\frac{\Delta i}{\Delta t}$$

$$\text{cemf} = -10 \times \frac{(2-1)}{1}$$

3. Solve to obtain cemf $= -10$ volts.

Example 5

If the inductance is 1 H and the current change from 1 ampere to 2 amperes takes place in ⅒ of a second, what would be the cemf?

1. Find the formula with the proper symbols for what is given and what is needed to be found.
2. Substitute the values known into the formula:

$$\text{cemf} = -L\frac{\Delta i}{\Delta t}$$

$$\text{cemf} = -1 \times \frac{(2-1)}{0.1}$$

3. Solve to obtain cemf $= -10$ volts.

From these examples it may be seen that a high value of opposition to the flow of current may be obtained either by increasing the inductance or the speed of the change of current in a circuit, or both. Thus, low-frequency a.c. circuits, because of the slow speed of change of the current, generally employ high values of inductance (with iron cores) to obtain a high cemf. High-frequency a.c. circuits, because of the great speed of change of the current, often may generate sufficient cemf with small air-core inductances. The following table illustrates the rise in cemf as the rate of change of the current increases:

L (henrys)	Δi (amperes)	Δt (seconds)	cemf (volts)
1	1	1	−1
1	1	0.5	−2
1	1	0.25	−4
1	1	0.1	−10
1	1	0.05	−20
1	1	0.02	−50
1	1	0.01	−100

From the table it is apparent that if a change of 1 ampere were to take place instantaneously, that is, if $\Delta t = 0$, the induced voltage, e, would become infinitely large. This would violate Kirchhoff's first law, which states that at any instant the applied voltage must equal the sum of the voltage drops around the circuit. Certainly, if $\Delta t = 0$, the voltage drop across the inductance would be greater than the applied voltage could be. And, by extension, it may be seen that at any instant, no matter how fast the change of current or how great the value of inductance, the induced voltage cannot be greater than the applied voltage. On the other hand, if there were no change of current, that is, if Δt were equal to infinity, the circuit would be a d.c. circuit, and e would be zero.

PROBLEMS

1. Find the missing value in the table below.

	L (henrys)	Δi (amperes)	Δt (seconds)	cemf (volts)
a.	1	2	3	
b.	2	2	3	
c.	3	2	3	
d.	1	1	0.1	
e.	1	1	0.01	
f.	1	1	0.001	
g.	1	1	0.0001	
h.	5	1	5.0	
i.	10	2	0.5	
j.	8	4	0.25	

2. Find the missing value in the table below.

	L (henrys)	Δi (amperes)	Δt (seconds)	cemf (volts)
a.		1	1	−1
b.		1	0.5	−2
c.		1	0.25	−4
d.	1	1		−10
e.	1	1	0.05	
f.	2	1	0.05	
g.	2	1	0.01	
h.	4	2	0.001	
i.	0.5	1	0.1	
j.	1	1	0.001	

INDUCTANCE IN SERIES AND PARALLEL

The total inductance of a circuit containing more than one inductance connected in *series* is calculated in the same manner as that used for resistors connected in series. The total inductance is the sum of the separate inductances:

$$L_T = L_1 + L_2 + L_3 + \ldots$$

This formula holds true only for circuits in which the inductances are shielded from each other, or so situated in relation to each other that no mutual inductance exists between them. See Fig. 5-10 for an illustration of mutual inductance.

If two inductors are connected in series and so arranged that the flux lines link each other, the resultant total inductance must include the mutual inductance present. That means the following formula must be used:

$$L_T = L_1 + L_2 \pm 2M$$

Both plus and minus signs are used in this general expression be-

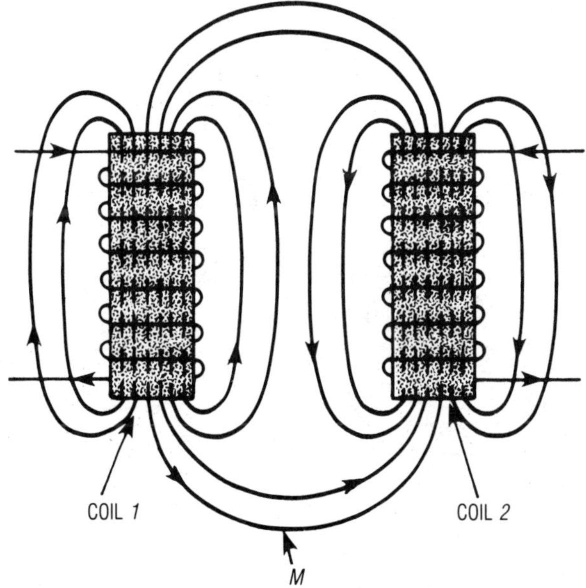

Fig. 5-10

cause the coils may be so placed that the induced voltages either aid or oppose each other. This increases or decreases the total inductance. Figure 5-11 shows the two inductances connected in series aiding, for which the formula is:

$$L_T = L_1 + L_2 + 2M$$

The lower half of Fig. 5-11 shows the same inductances connected in series opposing. Therefore the formula reads:

$$L_T = L_1 + L_2 - 2M$$

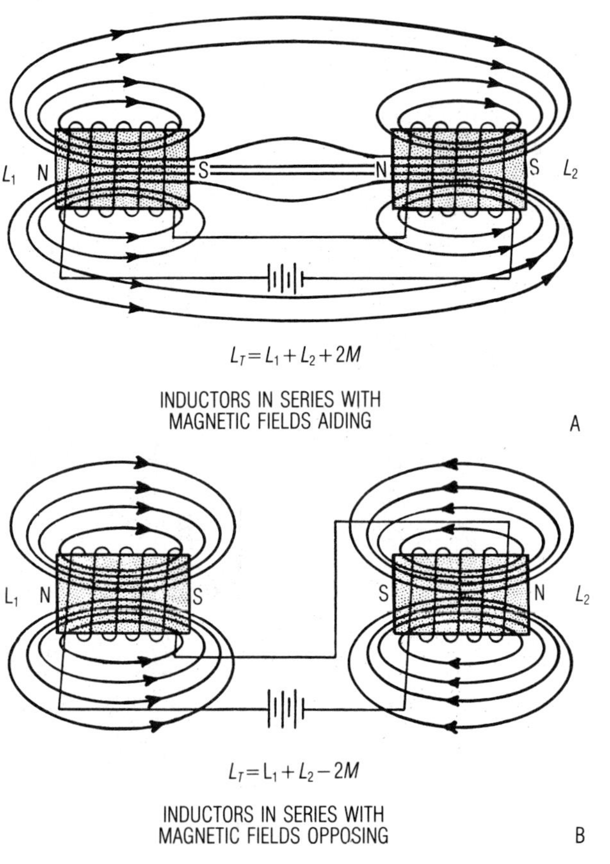

$L_T = L_1 + L_2 + 2M$

INDUCTORS IN SERIES WITH MAGNETIC FIELDS AIDING

A

$L_T = L_1 + L_2 - 2M$

INDUCTORS IN SERIES WITH MAGNETIC FIELDS OPPOSING

B

Fig. 5-11

If the coils are so arranged that one can be rotated relative to the other, changing the coefficient of coupling *K*, the total inductance in the circuit can be varied. This principle was used quite a lot in the early days of radio in what was called the variocoupler and the variometer.

Inductances in Parallel

The total inductance of a circuit containing more than one inductance connected in parallel is calculated in the same manner as that used for resistors in parallel. The total inductance is the reciprocal of the sum of the reciprocals of the individual inductances (provided the coils are shielded from one another and *M* equals zero). The formula then would read:

$$L_T = \frac{1}{\frac{1}{L_1} + \frac{1}{L_2} + \frac{1}{L_3}} + \cdots$$

If the coils are of equal inductance, total inductance may be obtained immediately by dividing the inductance of one coil by the number of coils in the circuit. Thus, four 2-H inductors total ½ H when connected in parallel.

For any two inductances, the simplified formula used with resistances may also be used:

$$L_T = \frac{L_1 \times L_2}{L_1 + L_2}$$

These formulas hold true only when the coils are shielded from one another and there is no mutual inductance. Any mutual inductance between coils in parallel has a tendency to reduce the total inductance. Thus, there can be no gain in total inductance and little practical use for unshielded inductances in parallel.

Example 6

What is the total inductance of a circuit with three coils in series, if the coils have 4 H, 2 H, and 1 H, and no mutual inductances exists among them?

1. Find the formula to use:

$$L_T = L_1 + L_2 + L_3$$

2. Substitute the values into the formula:

$$L_T = 4 + 2 + 1 = 7 \text{ H}$$

Example 7

What is the total inductance of a circuit if it has four 5-H coils in series aiding with 4 H of mutual inductance?

1. Find the formula to use:

$$L_T = L_1 + L_2 + L_3 + L_4 + 2M$$

2. Substitute in the formula and solve:

$$L_T = 5 + 5 + 5 + 5 + 2(4)$$
$$L_T = 20 + 8 = 28 \text{ H}$$

Example 8

What is the total inductance of a parallel circuit with coils connected in parallel, if the coils are 5 H, 4 H, 1 H, and 0.5 H? Assume no mutual inductance.

1. Find the correct formula:

$$L_T = \frac{1}{\frac{1}{L_1} + \frac{1}{L_2} + \frac{1}{L_3} + \frac{1}{L_4}}$$

2. Substitute the values:

$$L_T = \frac{1}{\frac{1}{5} + \frac{1}{4} + \frac{1}{1} + \frac{1}{0.5}}$$

3. Solve, using the calculator to obtain $L_T = 0.28986$ H. As you can see, the total is less than the smallest inductor.

Mutual Inductance

Mutual inductance means there is an interaction of the coils because of their physical placement with respect to one another. This is taken into consideration in the formula:

$$M = \frac{L_A - L_B}{4}$$

M is mutual inductance and is measured in henrys
L_A represents the total inductance of two coils (L_1 and L_2 when their fields interact to produce an aiding effect)
L_B represents the total inductance of two coils (L_1 and L_2 when their fields interact to produce an opposing, or canceling, effect)

Example 9

If you have two coils of 5 and 10 H interacting in series *opposing*, what would be the mutual inductance?

1. Find the proper formula and plug in the values.

$$M = \frac{(10+5)-(10-5)}{4}$$

2. Solve:

$$M = \frac{15-5}{5}$$

$$M = \frac{10}{4} = 2.5 \text{ H}$$

Coupled Inductances

In parallel circuits with inductances, if the fields are *aiding*, the formulas have to be adjusted for the effects of mutual inductance. Therefore, the formula would look like this:

$$\frac{1}{L_T} = \frac{1}{L_1 + M} + \frac{1}{L_2 + M}$$

Or:

$$L_T = \frac{1}{\dfrac{1}{L_1 + M} + \dfrac{1}{L_2 + M}}$$

Parallel inductances with their fields *opposing* require a few changes in the formula:

$$\frac{1}{L_T} = \frac{1}{L_1 - M} + \frac{1}{L_2 - M}$$

Or:

$$L_T = \frac{1}{\frac{1}{L_1 - M} + \frac{1}{L_2 - M}}$$

Series inductances with their fields *opposing* require some changes also. The following is the formula:

$$L_T = L_1 + L_2 - 2M$$

PROBLEMS

1. What is the total inductance of the coils mentioned if they are connected in series with no mutual inductance and are in series aiding? (All inductances are in henrys.)

	L_T	L_1	L_2	L_3	L_4
a.		10	5	2	1
b.		1	3	2	5
c.		1	10	2	3
d.		0.01	0.02	2	3
e.		0.1	0.15	0.3	0.125

2. What is the total inductance of the coils mentioned if they are connected in series *aiding* with the mutual inductance given?

	L_T	L_1	L_2	M
a.		1	1	2
b.		3	5	1
c.		4	6	3
d.		5	7	1
e.		0.5	0.3	1

3. What is the total inductance of the coils connected in series *opposing* with the mutual inductance given?

	L_T	L_1	L_2	M
a.		1	1	2
b.		3	5	1
c.		4	5	4
d.		4	3	1
e.		0.1	0.2	0.15

4. What is the total inductance if the following coils are connected in *parallel* with *no mutual inductance*?

	L_T	L_1	L_2	L_3	L_4
a.		1	1	1	1
b.		2	2	2	2
c.		2	4	6	8
d.		1.0	0.25	0.5	2.0
e.		1	2	3	4

5. What is the total inductance if the following coils are connected in *parallel with a mutual inductance* as indicated?

	L_T	L_1	L_2	aiding	M
a.		2	4		2
b.		1	4		1
c.		2	6		2
d.		3	4		1
e.		1	3		0.5

6. What is the total inductance if the following coils are connected in *parallel with a mutual inductance* as indicated?

	L_T	L_1	L_2	opposing	M
a.		10	5		4
b.		5	5		2
c.		1	2		0.5
d.		5	7		4
e.		6	10		4

TIME CONSTANT IN AN RL CIRCUIT

Inductance introduces a delay in time between the applied voltage and the current produced by this voltage in any circuit in which a change of current occurs. In a.c. sine-wave circuits, the current is a sine wave also, following the applied voltage by the phase angle θ. In such a circuit, θ depends on the ratio $\frac{X_L}{R}$—that is, on the frequency, inductance, and resistance of the circuit. In d.c. circuits (or circuits using pulses of direct current), however, the time lag resulting from the starting and stopping of the voltage does not depend on the frequency, since frequency is zero, but on the inductance and the resistance of the circuit.

When a direct current is applied to a pure resistance, the current is said to rise immediately to its $\frac{E}{R}$ value $(I = \frac{E}{R})$. But if an inductance is added to the circuit, the current is held back by the counter emf and does not rise to its $\frac{E}{R}$ value immediately. Figure 5-12 shows the graph of the time constant or rise of the current in a delayed fashion. This is in a circuit with direct current and a resistance and inductance. An RL circuit is one with resistance (R) and inductance (L). This curve is an exponential curve. That is, the current rises rapidly at first and then gradually tapers off to its

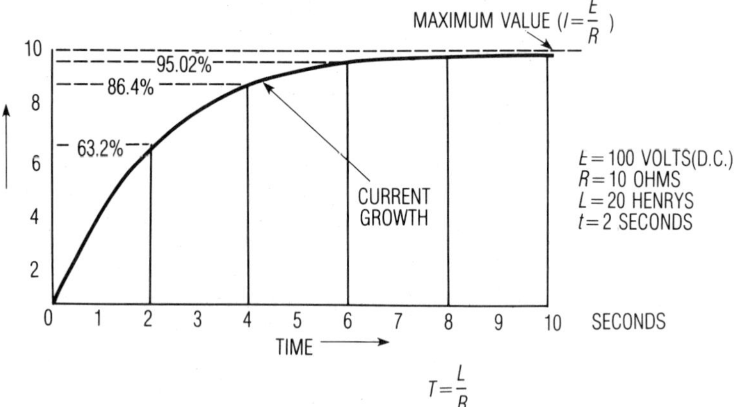

Fig. 5-12

maximum value, thus describing the characteristic curve shown. An analysis of this curve reveals that, in regular units of time, the current *rises in decreasing amounts*. Thus, in the first unit of time, current rises to 63.2% of the maximum possible. That means 36.8% remains to be reached before maximum. So the next unit of time takes 63.2% of the 36.8% to produce 86.4% of maximum—and in like manner for each succeeding unit of time. Theoretically, in such progression the current never reaches maximum because there is always a remainder, but for practical purposes current is considered at a maximum value after five time constants or units of time.

Table 5-1. Time Constants (in %)

1.	63.21205589
2.	86.46647168
3.	95.02129316
4.	98.16843611
5.	99.32620530
6.	99.75212478
7.	99.90881180
8.	99.96645374
9.	99.98765902
10.	99.99546001

As you can see from Table 5-1, the time constants do not come out in even numbers anywhere. Therefore, 63.2 and 86.5 are usually used. You can also see from the table that the 99% level is reached after five time constants; after that the change is insignificant. It was taken out to ten places here to show how you never reach the maximum value because there is always 63.2% of what is remaining.

Decay

The delay in having the current fall back to zero once the circuit has been deenergized is called *decay*. The curve in Fig. 5-13 shows the decay of current or the delay in getting back to zero. Theoretically, it will never reach zero, but practically, we drop off the inductive effect after five time constants. Note how the current falls rapidly at first and then tapers gradually to a minimum. In the first unit of time it falls 63.2% from its maximum $\frac{E}{R}$ value, or to 36.8%

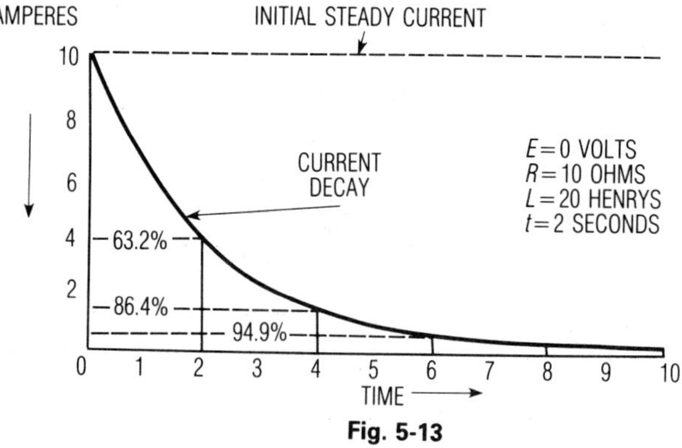

Fig. 5-13

of its maximum value. In the next unit of time it falls an additional 63.2% of the remainder. This process continues, as described above, until, after five units of time, the current is considered to be at zero.

This recurring 63.2% of the maximum rise or fall of current in a fixed unit of time is called the *time constant* of a circuit. The more inductance in a circuit, the longer the unit of time required to reach this initial 63.2% of the $\frac{E}{R}$ value. And the more resistance in a circuit, the shorter the time required to reach this value, since the greater the resistance of an RL circuit, the less effect the inductance has on the time constant. Therefore, you can see that the time constant is equal to inductance divided by resistance, or:

$$t\ (\text{time constant}) = \frac{L}{R}$$

L is in henrys
R is in ohms
t is in seconds

Example 10

What is the time constant of an inductor of 10 H in a circuit with 100 ohms resistance?

1. Which formula will you use? In this case it is the only one given:

$$t = \frac{L}{R}$$

2. Substitute the values in the formula to obtain:

$$t=\frac{10}{100}$$

3. $t=0.1$ second

Example 11

What is the current in a 100-volt d.c. circuit with an inductor and resistor if the inductance is 5 H and the resistance is 100 ohms?

1. What do you know and what do you need to know?
2. You know: $L=5$ H, $E=100$ V d.c., and $R=100$ ohms.
3. If you have these items given, the current will be:

$$I=\frac{E}{R}$$

$$I=\frac{100}{100}=1 \text{ ampere}$$

4. The problem did not ask you to find at what time the current reaches 1 ampere, but you know it takes a minimum of five time constants to rise to 1 ampere. The time constant in this case is not called for, but if you did want to find it, you would simply divide L by R, or 5 divided by $100=.05$ second, meaning it would take 0.25 second for it to reach 0.9932620530 ampere. ($0.05\times 5=0.25$ and the 0.9932620530 ampere is current after 5 time constants.)

Example 12

What would be the current in a circuit, after two time constants, with a resistor and inductor if it had 100 volts d.c. applied and the inductor is 0.5 H and the resistor is 100 ohms?

1. What do you know?
2. Known: $E=100$ V d.c., $R=100$ ohms, and $L=0.5$ H.
3. What do you want to know?
4. Unknown: time constant and second time constant value.
5. Select the formula that will give you the unknown time constant:

$$t=\frac{L}{R}$$

6. Substitute the values in the formula to obtain:

$$t=\frac{0.5}{100}=0.005 \text{ second}$$

or 5 milliseconds for one time constant.

7. The second time constant would be 86.5% of the maximum current and 10 milliseconds.

8.

$$I=\frac{E}{R}$$

$$I=\frac{100}{100}=1 \text{ ampere}$$

9. At the end of the second time constant the current in the circuit would be: $1 \times .865$ or 0.865 ampere.

PROBLEMS

1. Fill in the missing values:

	t (seconds)	*R* (ohms)	*L* (henrys)
a.		1000	1
b.	1		10
c.	2	10	
d.		1000	5
e.		100	0.5
f.		1000	0.12
g.	0.125	1000	
h.		4700	0.3
i.	3	10	
j.		10,000	0.15

2. What is the time constant of an RL circuit composed of an inductor of 10 H and a resistance of 500 ohms?
3. What is the current in a circuit composed of a source of 10

volts, an inductor of 6 H, and a total circuit resistance of 30 ohms at the end of 1 second?

4. What is the current in a circuit composed of a source of 10 volts, an inductor of 10 H, and a total circuit resistance of 100 ohms at the end of 0.5 second?
5. What is the time constant for an inductor-resistor circuit when the resistance is 4,700 ohms and the inductance is 100 microhenrys (μH)?
6. What is the time constant for an inductor-resistor circuit when the resistance is 47,000 ohms and the inductance is 47 mH?
7. What is the inductance of a circuit with a choke (inductor) and a resistance in series if the time constant is 1 second and the resistance is 10 ohms?
8. What is the resistance of a circuit with a choke and a resistance in series if the time constant is 3 seconds and the choke has 10 H of inductance?
9. A choke with an internal resistance of 50 ohms has an inductance of 5 H. What is its time constant?
10. What is the time constant of a circuit with 100 mH inductance and 1,000 ohms resistance?

INDUCTIVE REACTANCE

The opposition offered to a specific change of current by an inductance is measured at any given instant in terms of counter emf (the voltage bucking the applied voltage). In d.c. circuits, any opposition to current flow is termed resistance and measured in ohms. In a.c. circuits, it is convenient to measure inductive opposition in ohms rather than in counter emf or in volts. This type of a.c. opposition is called *inductive reactance* and is assigned the symbol X_L to distinguish it from a d.c. resistance.

Inductive reactance depends directly on the size of the inductance and on the rate of change of the current. In an a.c. circuit, the rate of change of the current is governed by the angular velocity of the applied voltage, and this velocity is measured in *radians*.

Since each hertz is 2 radians, the angular velocity of any sine wave is equal to this factor multiplied by the frequency. Therefore, the inductive reactance of any circuit is equal to the inductance multiplied by the angular velocity ($2\pi f$), or:

$$X_L = 2\pi f L$$

X_L = inductive ohms measured in ohms
$2\pi = 6.283185308$
f = frequency in hertz
L = inductance in henrys

From this formula it may be seen that the higher the frequency, or the greater the inductance, the greater the inductive reactance (X_L). A of Fig. 5-14 is a graph of X_L plotted against the frequency. The graphs in B of Fig. 5-14 is the inductive reactance plotted against inductance. It illustrates that the reactance increases directly, or linearly, with the frequency and inductance. The greater the frequency for any L, or the greater the inductance for any f, the greater the inductive reactance of the circuit.

At the high frequencies used in electronic communications, even small amounts of inductance may offer very great inductive reactance. C of Fig. 5-14 is a graph of current in an inductive circuit plotted against frequency. It may be said, then, from this that at high frequencies an inductance tends to act as an open circuit, since very little current flows in the circuit. At low frequencies, an inductance tends to act as a simple conductor—a short circuit—since high current flows. Thus, inductances designed for use at low frequencies are generally large iron core inductances. They are measured in henrys. Those used at high frequencies are generally small air-core inductances that are measured in millihenrys (0.001 H) or in microhenrys (0.000001 H).

Example 13

What is the opposition offered by a 20-H inductance to a 120-volt, 60-Hz a.c. circuit?

1. Determine the formula to use, in this case $X_L = 2\pi f L$.
2. Substitute the values and solve.
3. $X_L = 6.283185308 \times 60 \times 20 = 7{,}539.82237$ ohms

4. Now that you have the X_L, you can find the current by using Ohm's law for a.c. circuits:

$$I = \frac{E}{X_L}$$

5. Substitute the values in Ohm's law:

$$I = \frac{120}{7{,}539.82237}$$

$$I = 0.0159154943 \text{ ampere}$$

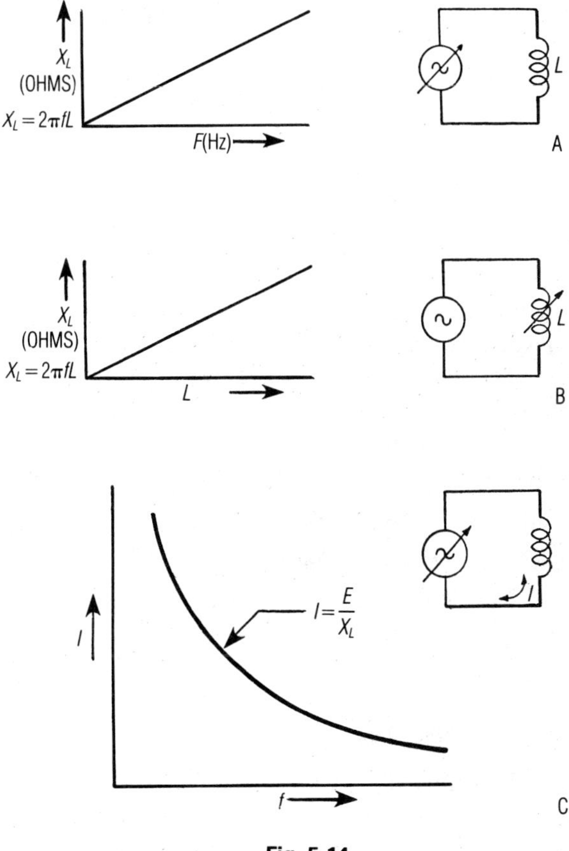

Fig. 5-14

Example 14

The frequency of a 120-volt, 20-H circuit is 400 Hz. What is the inductive reactance?

1. Determine the formula to use to find X_L, in this case $X_L = 2\pi fL$.
2. Substitute the values and solve.
3. $X_L = 6.283185308 \times 400 \times 20 = 50{,}265.48243$ ohms
4. Now that you have the X_L, you can find the current by using Ohm's law for a.c. circuits:

$$I = \frac{E}{X_L}$$

5. Substitute the values in Ohm's law:

$$I = \frac{120}{50{,}265.48243}$$
$$I = 0.002387324146 \text{ ampere}$$

Example 15

A transmitter operating at 5 MHz has a small inductance of 2.5 mH. It is used to keep substantially all alternating current out of a certain part of the transmitter while at the same time allowing the passage of direct current. This type of inductance is called a choke, since inductive reactance at that frequency is quite high and has effectively choked off the a.c. currents. What is the X_L of this circuit?

1. Determine the formula to use, $X_L = 2\pi fL$.
2. Substitute the values and solve.
3. $X_L = 6.283185308 \times (5 \times 10^6) \times (2.5 \times 10^{-3})$
4. $X_L = 78{,}539.81365$ ohms

Finding a Different Unknown

Suppose you have the inductive reactance and the frequency and need the inductance. You can still find the missing values using the original X_L formula.

Example 16
What is the inductance of a coil if it has 1,884 ohms inductive reactance and a frequency of 60 H?

1. Determine the formula to use: $X_L = 2\pi fL$.
2. Substitute what you know into the formula, to obtain:

$$1{,}884 = 2 \times 3.141592654 \times 60 \times \mathrm{L}$$

3. $1{,}884 = 376.9911185L$
4. $\dfrac{1{,}884}{376.9911185} = L$
5. $L = 4.997465212$ or 5 H

Suppose you don't know what the frequency is but you do know the inductance and the power line frequency.

Example 17
What is the frequency applied to a circuit with an inductor of 10 H that produces an inductive reactance of 1,884 ohms?

1. Determine the formula to use: $X_L = 2\pi fL$.
2. Substitute what you have into the formula to obtain:

$$1{,}884 = 2 \times 3.141592654 \times 10 \times f$$

3. $1{,}884 = 62.83185308f$
4. $\dfrac{1{,}884}{62.83185308} = f$
5. $f = 29.98479127$ or 30 Hz for all practical purposes.

PROBLEMS

1. Determine the inductive reactance for the following circuits:

	X_L (ohms)	f (hertz)	L (henry)
a.		100	5
b.		200	5

	X_L (ohms)	f (hertz)	L (henry)
c.		300	5
d.		400	5
e.		500	5
f.		500	6
g.		500	7
h.		500	8
i.		500	9
j.		500	10

2. Determine the missing values in the table below:

	X_L (ohms)	f (hertz)	L (henry)
a.		60	5
b.		25	5
c.		60	10
d.		25	10
e.	301.5928948		0.12
f.	75.39822372	60	
g.	3769.911185	6000	
h.	34,577.51919		0.01
i.		50	5
j.		50	10

PEAK, AVERAGE, AND RMS VALUE OF VOLTAGE AND CURRENT

The *instantaneous value* of a sine wave of voltage is the value of the emf generated at any instant of time. Note that all instantaneous values of either alternating current or voltage are generally indicated by small letters, whereas all average, effective, and maximum values are indicated by capital letters. Fig. 5-15 shows this notation. At 0° the instantaneous value, e, of the emf is 0. Between 0° and 90° the value of e is increasing from 0 to the maximum value. At 90° the value of e is a maximum and is equal to the peak voltage, that is, to E (maximum), written E_m. Between 90° and 180° the value of e is decreasing from peak voltage to 0. In the next half-cycle, e in-

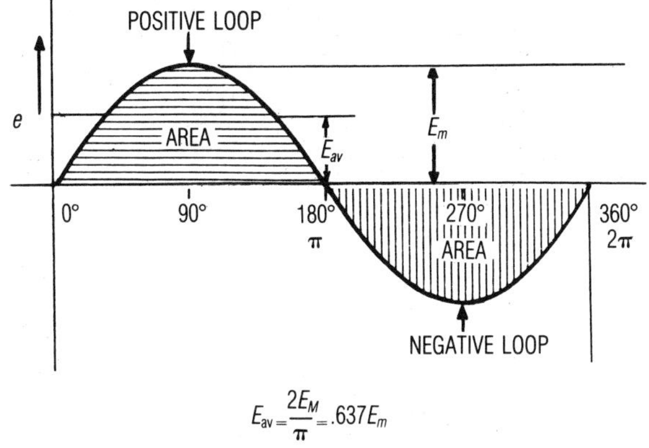

Fig. 5-15

creases and decreases in the same manner, but in the opposite direction. Thus, the instantaneous voltage varies constantly during one complete cycle of 360 electrical degrees. Peak voltage, E_m, may be defined, therefore, as the point of maximum instantaneous voltage. Peak voltage may be either the negative peak voltage or the positive peak voltage. For a *pure* sine wave of voltage, the positive peak is equal to the negative peak.

The average value of a complete cycle of sine wave voltage or current is zero. This is because the negative loop is equal and opposite to the positive loop. See Fig. 5-15. However, the term *average value*, when applied to an alternating current or voltage, is restricted to the average value of one loop, either the positive or negative. The positive loop of the sine wave is identical to the negative loop, and that means the area under each is the same.

The average value of a sine function is defined as the area under one loop divided by the base of the loop. The base of one loop is 180° or π radians in length. But in order to find the area of such an irregular surface as a sine loop, the figure must be broken into a series of small rectangles whose arcs may be easily determined. The sum of all these small areas will then be the area of the loop. Check out the negative loop in Fig. 5-15. In this manner it is found that the area of one loop of a sine curve is $2E_m$, where E_m is the maximum value of the voltage. Or, the area is equal to $2I_m$, where

I_m represents the maximum value of the current. The average value may then be defined by the following formula:

$$\text{Average value of the voltage} = \frac{2E_m(\text{area of loop})}{\pi(\text{length of base})}$$

or:

$$\frac{E_m}{\frac{\pi}{2}} = 0.637 \times E_m$$

$$\text{Average value of the current} = \frac{2I_m(\text{area of loop})}{\pi(\text{length of base})}$$

or:

$$\frac{E_m}{\frac{\pi}{2}} = 0.637 \times I_m$$

Effective or rms Value of Voltage or Current

In direct current you found that the value of the power dissipated as heat in a resistance is equal to I^2R. But when an alternating current flows through a resistance, the power dissipated does not remain constant over a full cycle (1 Hz), since the instantaneous current changes constantly with relation to time. However, the power absorbed by the resistor at any instant is equal to the square of the instantaneous current i multiplied by the resistance, or i^2R. At this point it should be noted that i^2 is always positive, even though i might be a negative value. This is because the square of a negative number is always positive and greater than zero.

Over a given period of time, such as one half-cycle, a certain amount of energy is delivered to the resistance in the form of heat. But it may be determined that a certain value of direct current flowing through the same resistance for the same amount of time produces the same heat dissipation as the alternating current. The value of this direct current is called the equivalent heating effect or the effective value of the alternating current. The effective current is defined as the area of one loop of the i^2 graph divided by π, or as the square root of the average value of i^2 over one loop, known as the root-mean-square, or rms. For this reason the effective value is often called the *rms value*. Figure 5-16 shows the curve of i^2. Keep in mind that all the loops are positive, since the squaring of i makes the negative loops positive in value. To find this effective, or rms, value of current, it is necessary to proceed in a manner similar to that used in determining the average value. The effective current then may be shown to be equal to the maximum current, I_m, divided by $\sqrt{2}$, and in like manner the effective voltage is equal to the peak voltage, E_m divided by $\sqrt{2}$. The effective value may then be expressed in the following formulas:

$$\text{rms Value of Current} = \frac{I_m}{\sqrt{2}} = \frac{I_m}{1.414} = 0.707 I_m$$

Therefore:

$$\text{rms} = 0.707 \times \text{maximum or peak current}$$

$$\text{rms Value of Voltage} = \frac{E_m}{\sqrt{2}} = \frac{E_m}{1.414} = 0.707 E_m$$

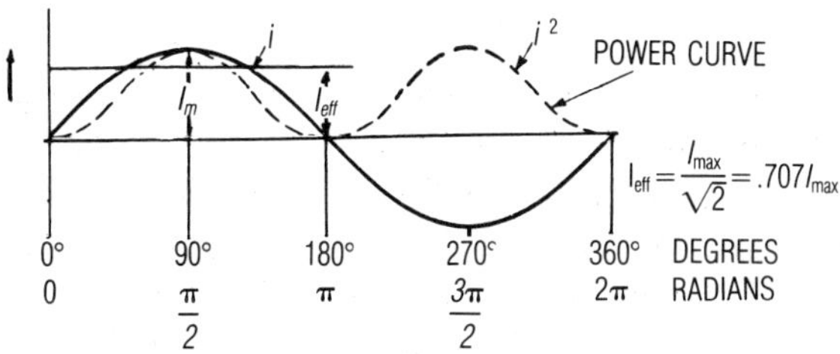

Fig. 5-16

Example 18

The voltage measured at an outlet in the home is 120 volts. Since this is the effective voltage, what is the peak voltage?

1. State the formula:

$$E_{\text{rms}} \times 1.414 = E_{\max}(\text{peak})$$

2. Substitute and solve.
3. $120 \times 1.414 = \text{peak}$
4. Peak voltage (maximum or $E_{\max}$) = 169.68 volts

You can *also* solve by:

1. $E_{max} = \dfrac{E_{rms}}{0.707}$
2. $E_{max} = \dfrac{120}{0.707}$
3. $E_{max} = 169.73$ volts

As you can see from this example, multiplying the effective voltage by 1.414 is the same as dividing the effective by 0.707.

Example 19

The voltage at the outlet of a home is 117 volts. Since this is the effective voltage, what is the average voltage?

1. State the formula:

$$E_{av} = E_m \times 0.637$$

   This means you have to find the maximum or peak voltage first to satisfy the formula. This can be done by:

2. $E_m = E_{rms} \times 1.414$
3. $E_m = 117 \times 1.414$
4. $E_m = 165.44$ volts
5. Now you can substitute in the average formula to find the average voltage:

$$E_{av} = 165.44 \times 0.637$$
$$E_{av} = 105.39 \text{ volts}$$

Example 20

An electric toaster uses an a.c. source to draw 4 amperes. This is the rms current. What is the maximum current value?

1. Select the proper formula:

   $I_{rms} = I_m \times 0.707$

2. $4 = I_m \times 0.707$

3. $I_m = \frac{4}{0.707}$

4. $I_m = 5.658$ amperes

Example 21

An electric iron uses an a.c. source to draw 4 amperes. This is the rms value of current. What is the average current?

1. Select the proper formula (keep in mind that 4A rms = 5.658 A max.):

$$I_{av} = I_m \times 0.637$$

2. $I_{av} = 5.658 \times 0.637$
3. $I_{av} = 3.604$ amperes

Conversion Formulas

In the examples just reviewed, you have seen that the peak voltage and the maximum current in an electric circuit are considerably in excess of what is generally the voltage or current given by the rms value. For this reason, circuits must be designed to withstand these values, even though they are instantaneous values and occur only twice each hertz or cycle. The following list of formulas, therefore, summarizes the findings and provides a useful means of converting one value to the other. The relationships among these values are also shown in Fig. 5-17.

$$E_{rms} = \frac{E_m}{1.414}$$

$$E_{rms} = E_m \times 0.707$$

$$E_{av} = \frac{E_m}{\frac{\pi}{2}}$$

$$E_{av} = E_m \times 0.637$$

$$E_m = E_{eff} \times 1.414$$

$$E_m = E_{av} \times \frac{\pi}{2}$$

Also:

$$E_{rms} = E_{av} \times 1.11$$
$$E_{av} = E_{rms} \times 0.9$$

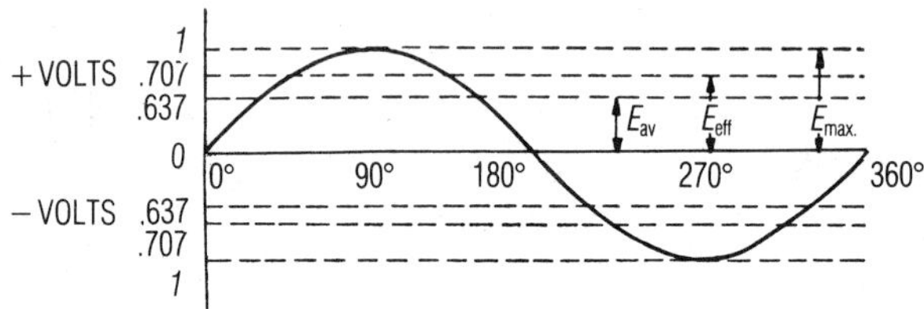

Fig. 5-17

PROBLEMS

1. Convert the rms voltages below to peak voltages.

	rms Voltage	Peak Voltage
a.	100 Volts	
b.	200 Volts	
c.	300 Volts	
d.	400 Volts	

2. Convert the peak voltages below to rms voltages.

	Peak Voltage	rms Voltage
a.	100 Volts	
b.	200 Volts	
c.	300 Volts	
d.	400 Volts	

3. What is the average current of the following rms currents?

	rms (A)	Average (A)
a.	10	
b.	5	
c.	1 mA	

CHAPTER 6

Alternating Current and Capacitance

CAPACITANCE

Capacitance is that property of an electric circuit that tends to oppose a change in voltage. The key words here are *change* and *voltage*. Change in voltage is opposed by the capacitor. This can be explained by comparing a rigid body and an elastic body. A force applied directly to a rigid mass meets with immediate opposition, but a force applied to an elastic body, such as a steel spring, meets little or no opposition and then an increasing opposition as the elastic body or spring is compressed or extended. Thus, it may be seen that the inertia of an elastic body produces an effect opposite to that of a rigid body and complementary to it. Furthermore, if an elastic material, such as a rubber band, is stretched and then fixed in position, or if a spring is compressed and fixed in position, the work done in stretching or compressing is stored indefinitely in the elastic and is returned when the rubber band or steel spring is released.

The counterpart of these mechanical effects characteristic of elastic bodies manifests itself in electrical circuits as the property of capacitance, which produces an effect opposite to that of inductance and complementary to it. Thus, if a voltage is applied to an inductance, opposition is immediate and there is a delay in current rise through it. But if a voltage is applied to a circuit containing capacitance, current flows at a maximum almost instantaneously, and then gradually falls to zero as opposition to it builds up. Furthermore, if the applied voltage is removed, the current caused by the capacitance of the circuit is stored indefinitely in that circuit and may be used at some later time.

These effects of capacitance appear particularly in a capacitor that, physically, is any two conductors separated by an insulating material. See Fig. 6-1. But capacitance, like inductance, is ever-present in all types of electrical circuits. Random or stray capacitive and inductive effects may be observed in any circuit, just as the most rigid bodies may be said to be elastic to some extent, and the most elastic bodies display some rigidity.

An examination of capacitance in a d.c. circuit, in which its effects are noticeably present only at the moments of opening or closing the circuit, reveals the basic storage action of a capacitor and its delayed reaction to an applied voltage. In A of Fig. 6-2, a simple capacitor is shown in series with a battery and a switch. The switch is open and the capacitor uncharged—that is, no difference in potential exists between the plates. At the moment of closing the switch, shown in B, plate *G* is at zero potential with relation to plate *H*, and both plates are at different potentials from the terminals of the battery. Therefore, the free electrons on plate *G* are attracted to the positive terminal of the battery, and those on plate *H* are repelled by the negative terminal of the battery. Then, since in a series circuit current is everywhere the same, the number of electrons flowing out of plate *G* equals the number flowing into plate *H*. Thus, plate *G* is positive because of a deficiency of electrons and plate *H* negative because of an excess of electrons. The capacitor

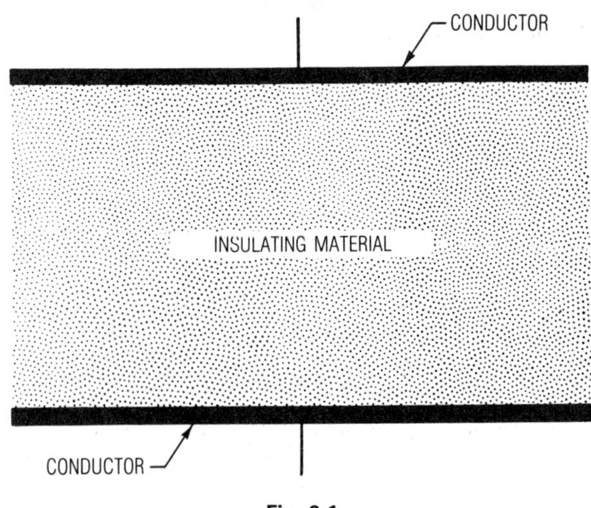

Fig. 6-1

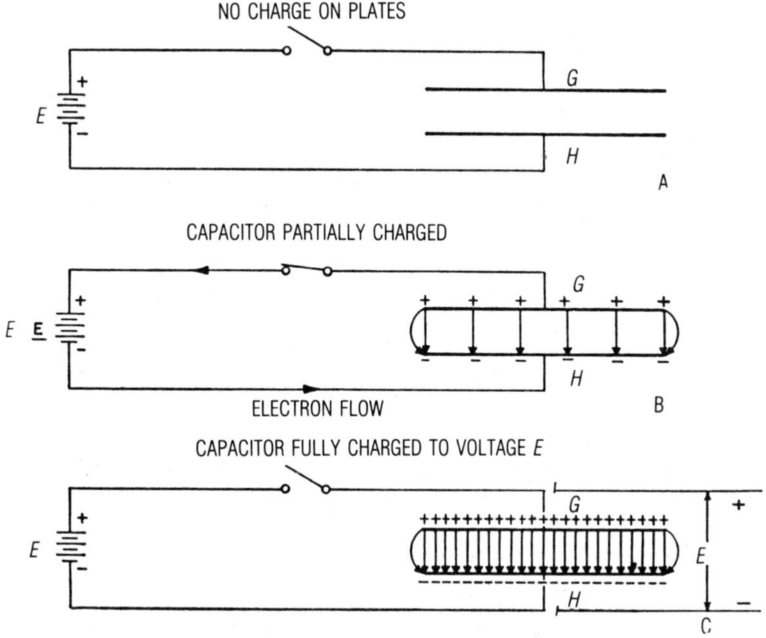

Fig. 6-2

is said to be fully charged when the difference in potential existing across the plates of the capacitor equals the battery voltage. In C, the switch is open and the capacitor remains charged, since there is no possible way in which the electrons on plate *H* can reach plate *G*. Thus, the capacitor stores the energy received from the battery and holds it until needed. If a conductor is placed across the plates of the capacitor, the electrons will flow from plate *H* to plate *G*, returning the capacitor to its original neutral uncharged condition.

Capacitance, then, is seen to be a kind of electrical inertia opposite in effect to inductance and similar to the property of elasticity in mechanical systems. Elasticity is defined as that property in nature that tends to oppose a change in force. In like manner, since a capacitor offers no immediate opposition or reaction to an applied voltage (current flows), and offers a maximum reaction when the applied voltage is removed (capacitor charged), a capacitor may be said to offer always a delayed reaction to voltage. Capacitance, then, may be defined as that property of an electric circuit that tends to oppose a change in voltage.

Measuring the Charge

In the analysis above of the basic storage action of a capacitor, it will be seen that the number of electrons entering and leaving the plates depends on the free electrons available and on the force applied—that is, on the physical size of the plates and on the difference in potential between the battery terminals. If the applied voltage is high, the forces of attraction and repulsion are great and the charge deposited on the plate of the capacitor is also great. Thus, experimentally it was discovered that for a given capacitor, the *ratio* between the amount of this charge and the voltage causing it is always *constant*. Also, since the electrostatic charge on one plate is equal and opposite to the other plate, the charge on one plate may be taken as the amount of this charge, which is designated as Q. Therefore, the ratio of the charge Q to the voltage E is taken to be a measure of capacitor action, which is called *capacity* and labeled C. Thus:

$$C=\frac{Q}{E}$$

The unit of measurement of C is the *farad* (F). The farad is defined as the capacity of a capacitor on one plate of which a charge of 1 coulomb (6.25×10^{18} electrons) is deposited by a difference in potential of 1 volt. This unit of measurement, however, is too large for practical circuits, and so the μF (microfarad), or one-millionth of a farad (10^{-6} F), is generally used. In communications even smaller units of measurement are used. In radio and television you will find the picofarad (pF), or one-millionth of a millionth of a farad (10^{-12} F).

Example 1

If a capacitor connected across a 10-volt battery shows a charge on one plate of one-thousandth of a coulomb, what is the size of the capacitor?

1. Examine the formula and see if you have the proper quantities to insert in the formula:

$$C = \frac{Q}{E}$$

2. Substitute the values you have given into the formula, to obtain:

$$C = \frac{0.001}{10} = 0.0001 \text{ F or } 100\ \mu\text{F}$$

Magnitude of the Charge

You can compute the magnitude of a charge on the capacitor by changing the formula to accommodate your needs. If $C = \frac{Q}{E}$, by using algebra you can see that $Q = CE$. So, if you want to find the magnitude of a charge, you can do it by multiplying the farads by the voltage.

Example 2

If you have a 100-μF capacitor and the voltage is increased to 1,000 volts, what would be the magnitude of the charge on the capacitor?

1. Examine the formula and see if you have the proper quantities to insert in the formula $Q = CE$.
2. Substitute the values in the formula, to obtain:

$$Q = 0.0001 \times 1{,}000 = 0.1 \text{ coulomb}$$

Voltage Across the Capacitor Plates

Since the capacity for a given capacitor is a constant, the voltage across the capacitor may be computed if the charge is doubled. Thus, $E = \frac{Q}{C}$ by simple algebra from the first formula given.

1. Examine the formula and identify the proper form.
2. Place the values you have into the formula and solve:

$$E = \frac{0.200}{0.0001} = 2{,}000 \text{ volts}$$

3. 0.2 comes from doubling the charge; 0.0001 comes from the capacity of 100 μF.

PROBLEMS

1. What size capacitor would you have under the following conditions:

	C (farads)	Q (coulombs)	E (volts)
a.		0.1	10
b.		0.1	100
c.		0.01	10
d.		0.01	100
e.		0.001	10
f.		0.001	100
g.		0.0001	10
h.		0.0001	100
i.		0.00001	10
j.		0.00001	100

2. What is the magnitude of the charge on the following capacitors when the voltage is as stated:

	C (microfarads)	Q (coulombs)	E (volts)
a.	100		10
b.	100		100
c.	100		1000
d.	10		10
e.	10		100
f.	500		6
g.	500		12
h.	1,000		15
i.	1,000		150
j.	5,000		10
k.	5,000		100
l.	10,000		10

3. Under the given conditions listed, what would be the charge in voltage across the plates?

	C (microfarads)	Q (coulombs)	E (volts)
a.	100	0.1	
b.	100	0.01	
c.	100	0.001	
d.	10	0.1	
e.	10	0.001	
f.	1	0.1	
g.	1	0.01	
h.	1	0.001	
i.	0.5	0.1	
j.	0.5	0.01	

DETERMINING CAPACITANCE

In the experimental determination of the formula for capacity, the charge on two concentric spheres in a vacuum was measured and compared to the voltage differential between the spheres. Since, however, this shape of capacitor is impractical, the parallel-plate capacitor, or some variation of it, is generally used. See Fig. 6-3.

The plates of this capacitor are conceived to be sections of the surfaces of the two spheres of infinite radius, since the surface of a sphere of infinite radius is a plane. Then, the charge Q on one plate is seen to be a part of the total charge on the sphere of infinite radius, a part proportional to the difference in potential between them, since the closer the plates are one to the other, the less force is necessary to set up a given charge; and the greater the distance between the plates, the greater the force necessary to maintain that charge. Thus, if the distance between the plates is zero (plates touching), the voltage E is zero; and if the distance were infinite, the voltage would be infinite. Therefore, for a parallel-plate capacitor of linear dimensions that are large in comparison with the distance between the plates, the formula $C = \frac{Q}{E}$ may be written as:

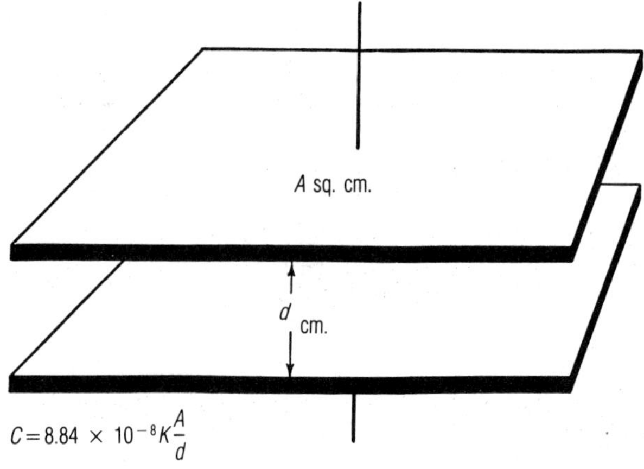

Fig. 6-3

$$C = \frac{A}{d}$$

A = area of one plate
d = distance between plates

However, in order to use this formula for values of capacity in microfarads and linear dimensions in centimeters, a conversion factor of 8.84×10^{-8} must be used to bring all quantities under the same system of units. Then:

$$C = 8.84 \times 10^{-8} \frac{A}{d}$$

A = the area of one plate in square centimeters
d = the distance between the plates in centimeters
C = the capacitance in microfarads

The same formula for values in inches and square inches becomes:

$$C = 22.45 \times 10^{-8} \frac{A}{d}$$

The above formulas hold for a parallel-plate capacitor in a vacuum—that is, with nothing between the plates, and vacuum as the *dielectric*, or insulating material. The introduction of some insulating substance between the plates intensifies the electrostatic field. This in turn produces a greater charge being held on the plates for the same difference in potential. Thus, by proper selection of this insulating material, capacitance may be increased since the ratio of Q to E is increased. Various materials have been tested experimentally to determine the ratio of this increase compared to a vacuum, which is taken as 1. This ratio is called the dielectric constant, K, of the material and is a factor independent of area or thickness, since in any single case it is a comparison with a vacuum of the same dimensions. The formulas then become:

$$C = 8.84 \times 10^{-8} K \frac{A}{d}$$

and:

$$C = 22.45 \times 10^{-8} K \frac{A}{d}$$

This formula reveals that the greater the dielectric constant, K, the greater will be the capacitance for a given set of dimensions.

The difference between the dielectric constant for air and for a vacuum is very small (air is 1.00059), and therefore, for practical purposes, air may be taken to be 1. Vacuum is also 1. Following are some typical values for K for the more common solid dielectrics:

Table 6-1. Dialectic Constants

Material	*K* (dielectric constant)
Air	1.0
Resin	2.5
Asbestos paper	2.7
Hard rubber	2.8
Dry paper	3.5
Isolantite	3.5
Common glass	4.2
Quartz	4.5
Mica	4.5–7.5
Porcelain	5.5
Flint glass	7.0
Crown glass	7–9
Ceramics	80–1200

Example 3

A parallel-plate air capacitor has plates 10×6 cm in dimensions and 0.1 cm apart. What is the capacity of the capacitor?

1. What is the formula to be used? The one with cm is to be used, since it uses the proper dimensions as given in the problem.
2. State the formula:

$$C = 8.84 \times 10^{-8} K \frac{A}{d}$$

3. Substitute the values in the formula:

$$C = \frac{8.84 \times 10^{-8} \times 1 \times 60}{0.1}$$

4. Reduce:

$$C = 8.84 \times 6 \times 10^{-6} = 53 \times 10^{-6}$$
$$C = 0.000053\ \mu\text{F or } 53\ \text{pF}$$

If a sheet of mica is placed between the plates, the capacity is increased. Since K for mica is approximately 6, the capacity would be: $C = 6 \times 53$ or 318 pF.

Example 4

If a parallel-plate capacitor of 0.02 μF uses paper 0.01 in. thick, what are the sq. in. dimensions of the plates?

1. Check for the proper formula to use. In this case it is:

$$C = 22.45 \times 10^{-8} K \frac{A}{d}$$

2. Substitute the values given into the formula. But before you can do that, you have to change the formula around to obtain the correct form for easy manipulation:

$$A = \frac{Cd \times 10^{8}}{22.45K}$$

3. Now you can substitute and solve:

$$A=\frac{0.02\times 0.01\times 10^{8}}{22.45\times 3.5}$$

$$A=\frac{2\times 10^{4}}{78.575}=254.5338848 \text{ sq. in. for plate area.}$$

If the plates are square, the dimension on a side is equal to the square root of 254.53389, or 15.954 in.

For larger capacities, the form of the basic parallel-plate capacitor is altered so that a series of plates and sheets of dielectric material may be stacked or piled one upon the other for greater compactness of construction. Figure 6-4 shows such a capacitor pile. Each group of alternate plates that are connected forms a plate of the capacitor, and both surfaces of the internal plates are used in calculations of total area.

Fig. 6-4

PROBLEMS

1. Find the capacity of the air capacitors below with the following dimensions:

	Capacity (microfarads)	Area (centimeters)	Number of Plates	Distance Apart (centimeters)
a.		10	10	0.1
b.		1	10	0.01
c.		100	1	0.001
d.		5	2	0.1
e.		7	6	0.25
f.		10	2	0.3
g.		100	2	0.5
h.		3	50	0.001
i.		4	25	0.0001
j.		10	5	0.01

2. What is the area of the plates of the following capacitors?

	Capacity (microfarads)	Area (sq. in.)	Dielectric Material	Dielectric Thickness (in.)
a.	0.0001		paper, dry	0.001
b.	0.001		paper, dry	0.001
c.	0.01		paper, dry	0.001
d.	0.1		paper, dry	0.001
e.	1		paper, dry	0.001
f.	0.047		paper, dry	0.01
g.	0.033		paper dry	0.01
h.	0.01		mica (4.5)	0.01
i.	0.033		glass, flint	0.001
j.	0.5		paper, asbestos	0.01

CAPACITORS IN SERIES AND PARALLEL

The total capacitance of a circuit containing more than one capacitor connected in series is calculated in the same manner as that for resistors in parallel. The total capacitance is the reciprocal of the sum of the reciprocal of the individual capacitors. Thus:

$$C_T = \frac{1}{\frac{1}{C_1} + \frac{1}{C_2} + \frac{1}{C_3} + \ldots}$$

An examination of this formula reveals that other generalizations true of *resistors in parallel* may be applied to *capacitors in series*.

1. The total capacitance is less than the capacity of the smallest capacitor.
2. If two capacitors are of equal value, total capacitance can be found by dividing one value by 2.
3. If three of the same value are in series, you divide the value of one by 3.
4. If any number of the same-size capacitors are connected in series, just divide the value of one by the number of capacitors in the string to produce the total capacitance.

5. For any two capacitors of different sizes, use the following:

$$C_T = \frac{C_1 \times C_2}{C_1 + C_2}$$

Take a look at Fig. 6-5A to see that two capacitors in series give a total capacitance less than either of those connected in series.

It will be noted that the addition of another capacitor in effect widens the distance between the plates while the area of the plates remains the same. Thus, total capacity is less since:

$$C = \frac{A}{d}$$

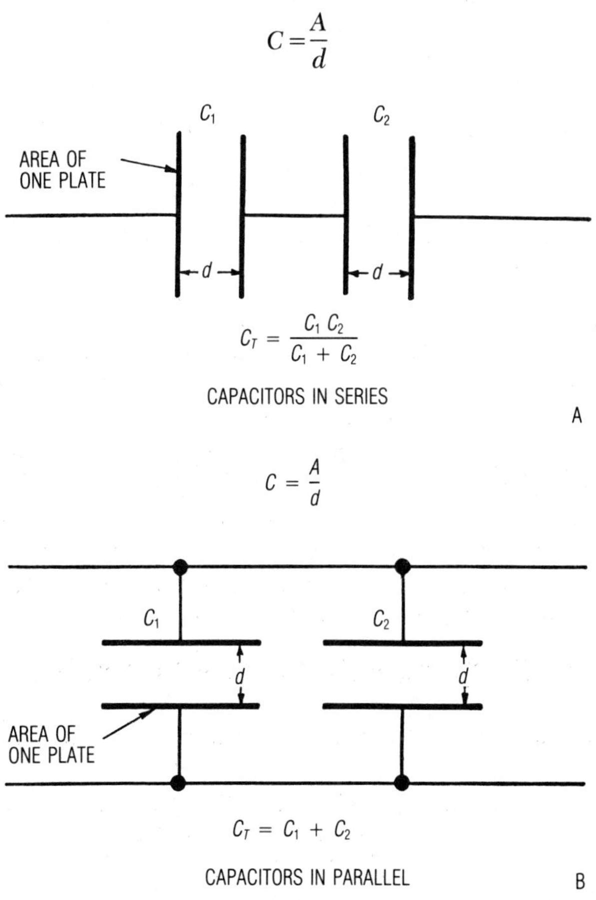

Fig. 6-5

Parallel Capacitors

On the other hand, the total capacitance of a circuit containing capacitors connected in *parallel* is calculated by the simple addition of the individual capacities, as with *resistances in series*. Thus:

$$C_T = C_1 + C_2 + C_3 + \ldots$$

In Fig. 6-5B two capacitors are connected in parallel. It will be noted that the addition of the second capacitor in effect increases the area of the plates without changing the distance between them. Thus, by the formula above, it is evident that total capacity increases and is always greater than the larger capacity.

Example 5

Two capacitors of 10 μF and 5 μF are placed in series. What is the total capacitance?

1. Use the formula and plug in the values.
2. $C_T = \dfrac{C_1 \times C_2}{C_1 + C_2}$ or $C_T = \dfrac{10 \times 5}{10 + 5} = \dfrac{50}{15} = 3.33\ \mu\text{F}$

Example 6

Three capacitors of 10 μF, 15 μF, and 30 μF are placed in series to make up a needed capacitance. What is the total capacitance?

1. Select the proper formula and fill in the values.
2. $\dfrac{1}{C_T} = \dfrac{1}{10} + \dfrac{1}{15} + \dfrac{1}{30}$
3. Find a common denominator. Or, use a calculator to obtain the reciprocals and add them. Then take the reciprocal of the sum of the reciprocals.
4. $$\frac{1}{C_T} = \frac{3 + 2 + 1}{30} = \frac{6}{30}$$
5. $$\frac{1}{C_T} = \frac{6}{30}$$

6. Invert for:

$$C_T = \frac{30}{6} = 5\ \mu F$$

Parallel Capacitors

Example 7

If you have connected two capacitors in parallel and each reads 10 μF, what is the total capacitance?

1. Remember the rule for capacitors in parallel. Just add the values to get the total.
2. 10 μF + 10 μF = 20 μF

The only problems associated with capacitors in parallel come when farads, microfarads, and picofarads are mixed and have to be converted to one of the three values and then added to produce whatever value you converted them to.

Problems

1. Find the total capacitance of the following *series* capacitors:

	C_T	C_1	C_2	C_3	C_4
a.		0.01 μF	0.01 μF	0.01 μF	0.01 μF
b.		0.01 μF	0.02 μF	0.03 μF	0.04 μF
c.		0.01 μF	10k pF	—	—
d.		0.1 μF	100k pF	—	—
e.		0.1 μF	10k pF	100k pF	—
f.		1 μF	10 μF	100 μF	1,000 μF
g.		10k pF	10k pF	10k pF	10k pF
h.		100k pF	100k pF	100k pF	0.1 μF
i.		100k pF	0.01 μF	0.00001 F	10k pF
j.		10 μF	10 μF	5 μF	5 μF

2. Find the total capacitance of the following *parallel* capacitors:

	C_T	C_1	C_2	C_3	C_4
a.		0.01 μF	0.01 μF	0.1 μF	0.1 μF
b.		10k pF	0.01 μF	1 μF	0.000001 F
c.		100k* pF	100k* pF	100k* pF	100k* pF
d.		0.1 μF	0.001 μF	10 μF	100 μF
e.		0.1 μF	100k pF	0.1 μF	100k pF
f.		0.047 μF	0.033 μF	0.05 μF	0.33 μF
g.		0.047 μF	47 pF	10 μF	30 μF
h.		10 μF	20 μF	30 μF	40 μF
i.		100 μF	100 μF	50 μF	25 μF
j.		1 μF	2 μF	40 μF	10k pF

* Keep in mind that k stands for kilo or 1,000.

WORKING VOLTAGE, DIRECT CURRENT (WVDC)

The capacity of a capacitor varies directly with the dielectric constant of the insulating material used. Therefore, for given dimensions and a given voltage, a capacitor using mica has six times the capacity of one using air. The charge for a given voltage is increased, since $Q = C \times E$.

The use of various dielectrics introduces an additional factor in the practical use of capacitors. Since the voltage appearing across the plates of a capacitor is often quite high, particularly in high-power communications circuits, the problem arises of breakdown or arcing over from one plate to the other. If the dielectric breaks down and becomes a conductor, the capacitor is useless, since it no longer can hold a charge and is said to be short-circuited. The voltage required to break down the dielectric varies with the material used and with its thickness. Thus, a great voltage would be required to break down a near-perfect vacuum, but a parallel-plate capacitor using a dielectric of air 0.001 in. thick breaks down at 80 volts. If the airspace is widened to 0.01 in., the breakdown voltage increases in proportion, to 800 volts.

But the capacity has been lessened since the plates are separated further. Other dielectric materials, however, not only increase capacity for given dimensions, but also increase the voltage necessary to cause a breakdown. Thus, mica 0.001 in. thick withstands 2,000 volts before breaking down, compared to the 80 volts it took for air. Not only that, it also increases the capacity six times over the capacitor using air as a dielectric. Mica is rarely used today as a dielectric; it has been replaced by ceramic materials with much higher voltages and dielectric constants.

Table 6-2 lists common substances used for dielectrics. Their

Table 6-2. Breakdown Voltages

Dielectric	*K*	Dielectric Strength (volts per 0.001 in.)
Air	1	80
Fiber	6.5	50
Bakelite	6	500
Glass	4.2	200
Mica	6.0	2000
Castor Oil	4.7	380
Paper		
Beeswax	3.1	1800
Paraffined	2.2	1200
Porcelain	5.5	750

approximate breakdown voltages per .001 in. of thickness is also given. Keep in mind, though, there is no relationship between dielectric *constant* and dielectric *strength* or breakdown voltage.

Another way of increasing the breakdown voltage is by putting two capacitors in series. You can increase it even more by putting three or more in series. As you put capacitors in series, you separate the plates further and increase the amount of voltage it takes to break them down.

Example 8

A capacitor of 10 μF with 100 WVDC is connected in series with another capacitor with 100 WVDC. What is the total capacity and the working voltage of the new capacitor?

1. To find the capacity in series you have to go to the formula used for that particular problem:

$$C_T = \frac{C_1 \times C_2}{C_1 + C_2}$$

$$C_T = \frac{10 \times 10}{10 + 10} = \frac{100}{20} = 5\ \mu\text{F}$$

2. Determine the working voltage by simply adding the WVDC ratings on the capacitors. In this case add the 100 + 100 to get 200 WVDC.
3. The answer is 5μF at 200 WVDC.

In the case of electrolytics, you may want to connect them back to back (− to − or + to +) to produce an a.c. electrolytic for use in an electric motor or crossover network for a high-fidelity stereo speaker enclosure.

In the case of the parallel capacitors, you have the WVDC for the smallest value of capacitor's working voltage. For example, you have two capacitors of 10 μF each, and they have working voltages of 100 WVDC for one and 10 WVDC for the other. You select the smaller one every time because in parallel they have their plates exposed to the applied voltage, and the weakest one (the one with the plates the closest) will arc over and thereby short out the whole group of capacitors in parallel.

Example 9

Two capacitors of 10 μF and 50 μF are connected in parallel. One capacitor has 100 WVDC and the other has 1,000 WVDC. What is the capacity and the working voltage for the combination?

1. Find the capacitance first. That can be done by simply adding the capacitor values.
2. 10 μF and 50 μF equal 60 μF for the total capacity.
3. Pick out the smallest working voltage, and that is the answer.
4. In this case you would have a capacitor combination with 60 μF at 100 WVDC.

Example 10

Three capacitors are connected in parallel. One has 10 μF at 10 WVDC, another has 100 μF at 5 WVDC, and the third has a capacity of 5 μF with a working voltage of 1.5 volts. What is the capacity and the working voltage of the new combination?

1. Add the capacitors to obtain the total: $10+100+5=115$ μF.

2. Select the smallest working voltage of the combination. In this case it is 1.5 WVDC.
3. That means the total combination would be 115 μF at 1.5 WVDC.

PROBLEMS

1. What is the capacitance and the working volts direct current (WVDC) of the following combinations of capacitors connected in *series*?

	Total Capacitance	WVDC	C_1	C_2	C_3
a.			10 μF @ 100 WVDC	20 μF @ 10 WVDC	5 μF @ 1.5 WVDC
b.			100 μF @ 1,000 WVDC	5 μF @ 100 WVDC	1 μF @ 15 WVDC
c.			100 μF @ 1,000 WVDC	100 μF @ 1,000 WVDC	100 μF @ 1,000 WVDC
d.			50 μF @ 10 WVDC	50 μF @ 10 WVDC	
e.			200 μF @ 1.5 WVDC	200 μF @ 100 WVDC	

2. What is the capacitance and the working volts direct current (WVDC) of the following combinations of capacitors connected in *parallel*?

	Total Capacitance	WVDC	C_1	C_2	C_3
a.			10 μF @ 100 WVDC	20 μF @ 10 WVDC	5 μF @ 1.5 WVDC
b.			100 μF @ 1,000 WVDC	5 μF @ 100 WVDC	1 μF @ 15 WVDC
c.			100 μF @ 1,000 WVDC	100 μF @ 1,000 WVDC	100 μF @ 1,000 WVDC

	Total Capacitance	WVDC	C_1	C_2	C_3
d.			50 μF @ 10 WVDC	50 μF @ 10 WVDC	
e.			200 μF @ 1.5 WVDC	200 μF @ 100 WVDC	

TIME CONSTANTS OF AN RC CIRCUIT

The property of capacitance, as has been shown, introduces a delay in time between the circuit current and the appearance of voltage in the circuit. In a.c. sine-wave circuits, the voltage is a sine wave also, following the current by the phase angle θ. In such a circuit, the phase angle depends on the ratio $\frac{X_C}{\mathrm{R}}$, that is, on $\frac{1}{2\pi fCR}$, the frequency, capacitance, and resistance of the circuit. In d.c. circuits (or circuits using pulses of d.c.), the time lag resulting from the starting and stopping of the current depends not on frequency, since frequency is zero, but on the capacitance and resistance of the circuit.

Thus, when a direct current is applied to an LR (inductor-resistor) circuit, the current is held back by the counter emf and does not rise immediately to its $\frac{E}{R}$ value. However, when a direct current is applied to an RC (resistor-capacitor) circuit, the current reaches its maximum $\frac{E}{R}$ value almost immediately and then gradually falls to zero as the counter emf offers greater and greater opposition. A of Fig. 6-6 shows the graph of current rise and fall in a d.c. circuit containing capacitance and resistance. Place B of Fig. 6-6 at the end of A and you have C. The voltage waveform is shown in D. It will be noticed that after the almost instantaneous rise of current to its maximum, the curve of falling current is an exponential curve and is the characteristic curve of current rise in an LR circuit turned inside out. Thus, the current falls rapidly at first and then gradually tapers off to its zero value. An analysis of this exponential curve reveals that in regular units of time, the current falls in decreasing

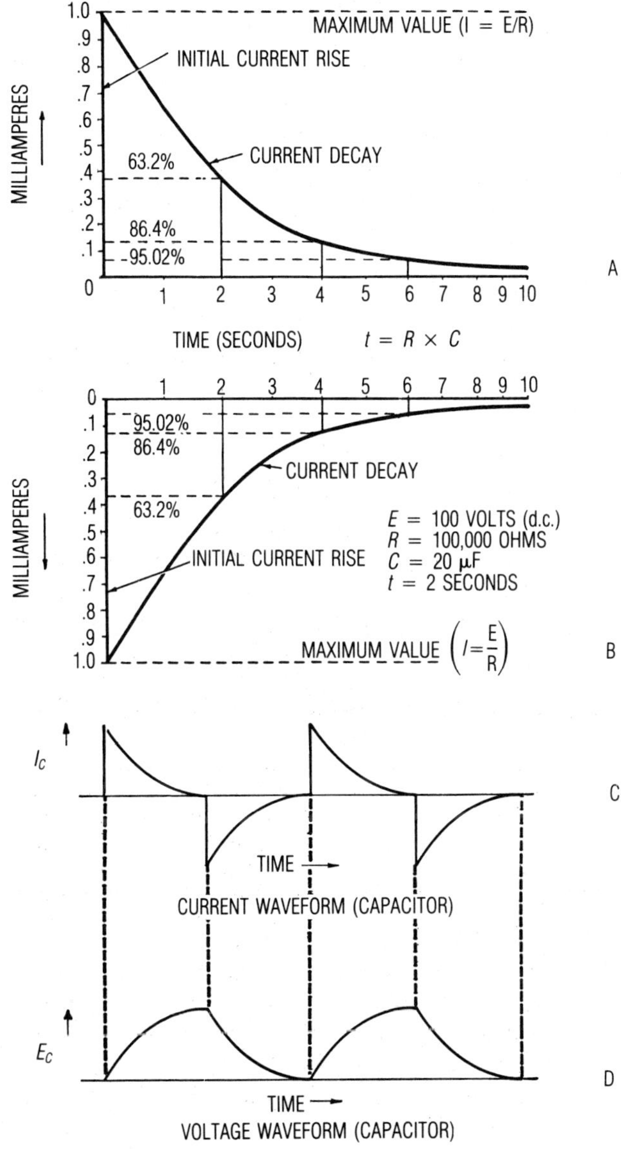
MAXIMUM VALUE (I = E/R)
INITIAL CURRENT RISE
CURRENT DECAY
63.2%
86.4%
95.02%
MILLIAMPERES
TIME (SECONDS)
t = R × C
A
CURRENT DECAY
INITIAL CURRENT RISE
E = 100 VOLTS (d.c.)
R = 100,000 OHMS
C = 20 μF
t = 2 SECONDS
MAXIMUM VALUE (I = E/R)
B
I_C
TIME
CURRENT WAVEFORM (CAPACITOR)
C
E_C
TIME
VOLTAGE WAVEFORM (CAPACITOR)
D

Fig. 6-6

amounts. Thus, in the first unit of time, current falls 63.2% of its maximum value—that is, to within 36.8% of the zero mark. In the second unit current falls 63.2% of the remaining 36.8% (86.4% from maximum), and in like manner for each succeeding unit of time. Theoretically, in such a progression current would never reach zero value, but for practical purposes current is considered to be zero and the capacitor fully charged after five units of time. In capacitor time constants:

$$T = R \times C$$

T is in seconds

R is in ohms

C is in farads

Ten time constants are shown below so you can see what happens after the fifth unit of time.

Table 6-3. Time Constants

Time Constants (in %)
1. 63.21205589
2. 86.46647168
3. 95.02129316
4. 98.16843611
5. 99.32620530
6. 99.75212478
7. 99.90881180
8. 99.96645374
9. 99.98765902
10. 99.99546001

On the other hand, if the d.c. source is removed and the capacitor is allowed to discharge, current flows again almost instantaneously at a maximum in the opposite direction, and then gradually returns to zero. B of Fig. 6-6 is the graph of the discharge current. This curve is the same as that described for the charging current, but turned upside down. In the first unit of time, current returns rapidly 63.2% of the remainder. This process continues as described above until, after five units of time, current is considered to be zero and the capacitor discharged.

This recurring 63.2% fall from maximum rise of current in a

fixed unit of time is called the *time constant* of an RC circuit. The more capacitance in a circuit, the longer time required to reach 63.2% of full charge. Also, the more resistance in an RC circuit, the longer the time required to reach the full charge and discharge, since the RC circuit tends to act as a resistive circuit and hold current to its $\frac{E}{R}$ maximum for a longer period. Therefore:

$$t \text{ (time constant)} = R \times C$$

If C is in microfarads (10^{-6}) and resistance is in megohms (10^6), then t (in seconds) is the time required to fall 63.2% from its $\frac{E}{R}$ maximum.

Example 11

If an RC circuit with 20 μF of capacitance and 100,000 ohms resistance is connected to 100 volts d.c., what would be the time constant?

1. You know the formula:

$$t = RC$$

2. Substitute in the values:

$$t = 0.1 \text{ megohm} \times 20\ \mu\text{F}$$
$$t = 2 \text{ seconds}$$

That means it will take the current 2 seconds to reach 63.2% of the maximum current and an additional 8 seconds to reach a full charge at five time constants. Or:

a. Full charge = 5 time constants
b. $t = 2$ seconds
c. Full charge $= 2 \times 5 = 10$ seconds

It would take 10 seconds for the capacitor to charge. In like fashion the current would drop to within 36.8% of its maximum in 2 seconds, but would take another 8 seconds to fully discharge.

Example 12

Since you already have the voltage at 100 volts d.c. and the resistance of 100,000 ohms given in the previous problem, what would be the maximum current in the circuit? What would be the current after 2 seconds?

1. The total current can be found by Ohm's law.

$$I=\frac{E}{R}$$

2. Substitute the values and solve:

$$I=\frac{100}{100{,}000}=1 \text{ milliampere or } 0.001 \text{ A}$$

3. The time constant was already found to be 2 seconds, so that means 63.2% of the current is present and the capacitor is charged by $63.3\% \times 0.001$ A at the end of 2 seconds. That means $0.632 \times 0.001 = 0.000632$ A or 0.632 milliampere.

Problems

1. What is the time constant for the following values?

	t (sec)	*C*	*R*
a.		0.01 μF	20 megs
b.		0.001 μF	2 megs
c.		0.0001 μF	200 *K*
d.		1.0 μF	470 *K*
e.		10.0 μF	470 *K*

2. If you have a voltage of 1,000 volts d.c. applied to an RC combination of 1 μF and 1 megohm, what would be the time constant?
3. What would be the total current or maximum current in Problem 2?
4. What would be the current after 1 second?
5. What would be the current after 2 seconds?
6. What would be the current after 4 seconds?
7. What would be the current after 5 seconds?
8. If you have a voltage of 1,000 volts d.c. applied to an RC

combination of 0.001 μF and 10 megs, how long would it take for the capacitor to fully charge?

9. How long would it take the capacitor in Problem 8 to reach 63.2% of its full charge?
10. What would be the current in the circuit in Problem 8 after 0.01 second?

CAPACITIVE REACTANCE

The conductance of a capacitor at any instant is expressed in terms of the current flowing in the circuit at that instant. The unit of measure of conductance, however, is the mho, and its symbol is G. The word *mho* is the word *ohm* spelled backwards. Conductance is the ease with which current flows in a circuit and is the inverse of resistance, the opposition to current flow. Thus, conductance is the reciprocal of resistance and resistance is the reciprocal of conductance. Then:

$$G\text{ (mhos)} = \frac{1}{R\text{ (ohms)}} \qquad \text{or} \qquad R\text{ (ohms)} = \frac{1}{G\text{ (mhos)}}$$

The conductance of a capacitor depends directly on the size of the capacitor and the rate of change of voltage. In an a.c. circuit, the rate of change of voltage is governed by the *angular velocity* of the applied voltage. This velocity is measured in *radians*. Since each hertz is 2π radians, the angular velocity of any sine wave is equal to this factor multiplied by the frequency (f). Therefore, the conductance of any capacitor is equal to the capacitance in farads multiplied by the angular velocity $2\pi f$, or:

$$G_C = 2\pi f C\text{(mhos)}$$

Then, since the reaction of an inductance to alternating current is measured in ohms, it is convenient to measure the reaction of a capacitor in ohms also. Therefore, inductive reactance X_L finds its counterpart in capacitive reactance, which is given the symbol X_C and measured in ohms. Capacitive reactance, then, is equal to the reciprocal of the conductance. Thus:

$$X_C = \frac{1}{G_C}$$

Substituting $2\pi fC$ for G_C:

$$X_C = \frac{1}{2\pi fC}$$

$2\pi = 6.283185308$
f = frequency in hertz
C = capacitance in farads
X_C = capacitive reactance in ohms

From this formula, it may be seen that the higher the frequency, or the greater the capacitance, the less the capacitive reactance.

At the very high frequencies used in communications, even small stray amounts of capacitance offer a path of such low reactance to the signal voltage that often a desired signal is short-circuited. At the other extreme, the separate d.c. circuits are effectively isolated from each other by either small or large capacitors, since either offers a path of maximum reactance to d.c.

NOTE: The latest changes in the units of measurement give conductance a basic unit of the siemen with an abbreviation of S. The conductance G is still used; only the unit of measurement for the G has changed.

PROBLEMS

1. What is the capacitive reactance of a circuit with 120 volts, 60 Hz applied to a capacitor with 10 μF capacitance?
2. What is the capacitive reactance of a capacitor of 0.01 μF in a circuit of 100 kHz?
3. What is the capacitance of a circuit if it has 1591.549431 ohms of reactance to 100 kHz?
4. What is the capacitance of a circuit with 1059.972981 ohms reactance to a 455 kHz signal?
5. What is the capacitance of a circuit that has 26.52582384 ohms reactance to 60 Hz?
6. Find the missing value in the following:

	X_C (Ω)	C (μF)	f
a.		0.001	100 kHz
b.		0.01	100 kHz
c.		0.1	100 kHz
d.		1.0	100 kHz
e.		10.0	100 kHz
f.		0.047	455 kHz
g.		0.047	55 kHz
h.		0.0033	1,600 kHz
i.		0.0033	1.000 MHz
j.		0.01	60 Hz

CURRENT IN CAPACITIVE CIRCUITS

A high current through a capacitor may be obtained either by increasing the capacitance or the speed of the change of the voltage across it, or both. For example, a 1-microfarad capacitor has 1,000 volts across the plates. If this voltage changes to 2,000 volts in 1 second, the current will be:

$$I = C\frac{\Delta e}{\Delta t}$$

This in turn is equal to:

$$1 \times 10^{-6} \times \frac{2000 - 1000}{1}$$

$$I = 1{,}000 \times 10^{-6} = 0.001 \text{ or } 1 \text{ milliampere}$$

If the change of voltage remains the same, but the capacitor unit is changed to 10 microfarads, then:

$$I = 0.010 \text{ or } 10 \text{ milliamperes}$$

If the capacitance remains, as in the first instance, at 1 microfarad, and the change of charge takes place in 0.1 second, then:

$$I = 0.010 = 10 \text{ milliamperes}$$

From the examples above, it may be seen that high-frequency a.c. circuits, because of the great speed of change of the voltage, pass sufficient current with small values of capacitance. It may also be seen that low-frequency a.c. circuits, because of the slow speed of change of the voltage, generally employ large values of capacitance in order to pass sufficient current. The following table illustrates, for a given capacitor and a fixed change in voltage, the rise in current as the rate of change of voltage increases:

Table 6-4. Current and Voltage Change

C (μF)	Δe (volts)	Δt (millisec)	I (milliamp)
1	100	100	1
1	100	50	2
1	100	25	4
1	100	10	10
1	100	5	20
1	100	1	100
1	100	0.5	200
1	100	0.2	500
1	100	0.125	800
1	100	0.100	1000

Δe = amount of change in voltage
Δt = amount of time required for change in voltage

From Table 6-4 it is apparent that if a change of voltage were to take place instantaneously, that is, if Δt were equal to infinity, the circuit is a d.c. circuit and the current would be zero. For any a.c. voltage, however, the rate of change depends on the number of hertz, or frequency. An increase in frequency means that the electrons must be taken from one plate and deposited on the other at a faster rate; then, since current is the rate at which charges are deposited, an increase in frequency brings about an increase in current. The current through a given capacitor, then, is directly proportional to frequency, and therefore the opposition offered by a capacitor is inversely proportional to the frequency. Thus, a capacitor reveals characteristics opposite in effect to an inductance.

Figure 6-7 shows how the current through an inductor and the voltage across a capacitor change with frequency, and also indicates the current in a capacitor and the voltage across an inductor in the straight line and how they vary with a change in frequency.

After this, you should be able to sum up the effects of a capacitor by the following statements:

1. A capacitor stores electrical energy and offers a delayed reaction to a change in voltage.
2. The ratio between the charge on a capacitor and the voltage causing it is a constant, called capacity:

$$C = \frac{Q}{E}$$

3. Different dielectrics show varying ratios of electrostatic permittivity as compared to air.
4. Different dielectrics show varying breakdown resistance to high voltages.
5. A capacitor is a conducting path to alternating current.
6. The current through a capacitor is determined by its capacity and the rate of change of voltage across it:

$$I = C\frac{\Delta e}{\Delta t}$$

7. A capacitor produces an effect opposite to that of an inductor, that is, the opposition to a.c. decreases with frequency.

NOTE: Problems and examples for this section will be found in the next section, "Coulomb's Law."

COULOMB'S LAW

In studying electrostatic charges, you found that the actual physical force between two charges (Q) separated by a distance depends on the magnitude of the charges and the square of the distance between them. Or:

$$F = \pm\frac{Q_1 Q_2}{d_2}$$

This is the expression for Coulomb's law of force between charges, and it indicates that the greater the charges and the smaller the distance between them, the greater the force between them. For the special case of a parallel-plate capacitor, Coulomb's law takes the form:

$$E = \frac{Q}{C}$$

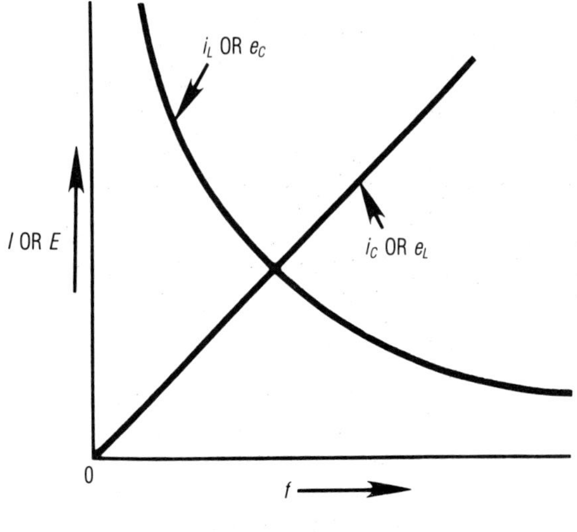

Fig. 6-7

As above, this formula indicates that the greater the charge and the smaller the capacity, the greater the voltage across the capacitor. For any given capacitor, then, it will be seen that the voltage across the capacitor is directly dependent on the magnitude of the charge Q. In an a.c. circuit, however, the charge Q on the capacitor is constantly changing in magnitude, and hence the voltage across the capacitor, which varies directly with the charge, is constantly changing in magnitude. This variation of charge, or flow of charge, is *current*, which is the *time rate of change of charge*. Hence:

$$I=\frac{Q}{t}$$

Q = the charge on one plate of the capacitor in coulombs
t = the time in seconds required to build up that charge
I = the current in amperes

Thus, if 0.2 coulomb of electrons is deposited in 1 second on the negative plate of a capacitor, the current in the circuit is 0.2 ampere. The magnitude of the voltage depends on the magnitude of the charge, but it should be noted that the magnitude of the current depends on the *rate of change* of the charge. However, since Q is equal to CE, the expression above may be written:

$$I = \frac{CE}{t}$$

Since the physical factors expressing the magnitude of the charge are contained under the capacitance, then C may be isolated and the rate of change of the charge expressed in terms of the rate of change of voltage. Then:

$$I = C\frac{E}{t}$$

Thus, the magnitude of the current through a capacitor is greater for a rapidly changing voltage (t is small) than for a slowly changing voltage (t is large). Therefore:

$$I = C\frac{\Delta e}{\Delta t}$$

Δe = the change in voltage in volts
Δt = the change in time, in seconds
C = the capacitance in farads
I = the current in amperes

The above formula reveals that the greater the capacity, or the faster the rate of change of the voltage, the greater the current in the circuit. Also, the farad may now be redefined as follows: *A farad is the capacitance of a circuit in which a voltage change of 1 volt per second causes a current flow of 1 ampere.*

Example 13

What is the current in the circuit if the capacitor is 100 μF and the change is 200 volts in one second?

1. Find the formula with the values given and fill in the formula:

$$I = C\frac{\Delta e}{\Delta t}$$

2. Substitute the values given in the problem to find the unknown:

$$I = 0.0001\ \frac{200}{1}$$

$$I = \frac{0.0001 \times 200}{1} = 0.02 \text{ ampere}$$

Example 14

The charge on one plate of the capacitor is 0.5 coulomb. The charge was deposited in 2 seconds. What is the current?

1. Find the correct formula:

$$I = \frac{Q}{t}$$

2. Substitute the values in the formula and solve:

$$I = \frac{0.5}{2} = 0.25 \text{ ampere}$$

Example 15

What is the voltage across the capacitor plates if the charge is 2 coulombs and the capacitor is 100 μF?

1. Find the correct formula:

$$E = \frac{Q}{C}$$

2. Substitute the values in the formula:

$$E = \frac{2}{0.0001} = 20{,}000 \text{ volts}$$

PROBLEMS

1. What is the current in the circuit if the capacitor is 10 μF and the change in voltage is 100 volts in one second?
2. What is the current in the circuit if the capacitor is 100 μF and the change in voltage is 10 volts in 2 seconds?
3. What is the current in the circuit if the capacitor is 1 μF and the change in voltage is 1,000 volts in 3 seconds?

4. What is the current in the circuit if the capacitor is 0.01 μF and the change in voltage is 300 volts in one second?
5. What is the current in the circuit if the capacitor is 0.001 μF and the change in voltage is 100 volts in 2 seconds?
6. The charge on one plate is 0.5 coulomb. The charge was deposited in 1 second. What is the current?
7. The charge on one plate is 0.25 coulomb. The charge was deposited in 2 seconds. What is the current in the circuit?
8. The charge on one plate is 0.1 coulomb. The charge was deposited in 2 seconds. What is the current?
9. The charge on one plate is 0.125 coulomb. The charge was deposited in 1 second. What is the current in the circuit?
10. The charge on one plate is 0.3 coulomb. The charge was deposited in 3 seconds. What is the current?
11. What is the voltage across the capacitor terminals if the capacitor is 10 μF and the charge is 1 coulomb?
12. What is the voltage across the capacitor terminals if the capacitor is 100 μF and the charge is 1 coulomb?
13. What is the voltage across the 100-μF capacitor terminals if the charge is 2 coulombs?
14. What is the voltage across the capacitor terminals if the capacitor is 5 μF and the charge is 0.5 coulomb?
15. What is the voltage across the capacitor terminals if the capacitor is 1 μF and the charge is 0.5 coulomb?

CHAPTER 7

Impedance and Phase Angles

IMPEDANCE

In the use of direct current we find the opposition to current flow to be resistance. In circuits which contain alternating current and resistors or some kind of resistance only, the total opposition is still resistance and is measured in ohms. However, when an inductor or capacitor is introduced in the circuit with alternating current, there is a reactive element which must be considered. We have already explored the ramifications of the capacitor in an a.c. circuit and found the opposition to current flow to be called capacitive reactance, X_C, and measured in ohms. We have also found that an inductor puts up an opposition to current flow that we refer to as inductive reactance, X_L, which is also measured in ohms. Once we put a capacitor and resistor, a capacitor and inductor, an inductor and resistor, or any combination of resistors, inductors, and capacitors in an a.c. circuit, we have another type of opposition to current flow which we refer to as *impedance*. Impedance impedes, or opposes, the current flow in an a.c. circuit. Keep in mind now that a resistor has only resistance, a capacitor has capacitive reactance, and an inductor has inductive reactance to oppose current flow. When we put any combination together they produce impedance. Impedance is represented by the letter Z. Impedance is also measured in terms of ohms. However, impedance is not measured by a meter; it is figured *mathematically*. That is where the formulas for impedance come in handy.

The following sections will give you an idea as to how important and involved the term *impedance* can be.

PHASE SHIFT

Before we undertake the problems associated with a.c. and inductance, we should have a good grasp on phase shift and what it is. Keep in mind that inductive reactance not only limits the current flowing in an a.c. circuit, but also tends to retard the building up and falling off of current. In A of Fig. 7-1, a sine wave of voltage is applied to a pure inductance. The current in the circuit also follows the form of the sine wave. But it is necessary to determine precisely the delay in time between the application of the maximum voltage and the moment the current reaches maximum value—that is, the phase shift between the voltage and current.

By Kirchhoff's first law it is known that the algebraic sum of the voltage drops about any closed circuit is equal to zero. That means the following formula is easily developed from that information:

$$e \text{ (applied)} + \text{cemf} = 0$$
$$e = -\text{cemf}$$
$$e = -\left(-L\frac{\Delta i}{\Delta t}\right)$$

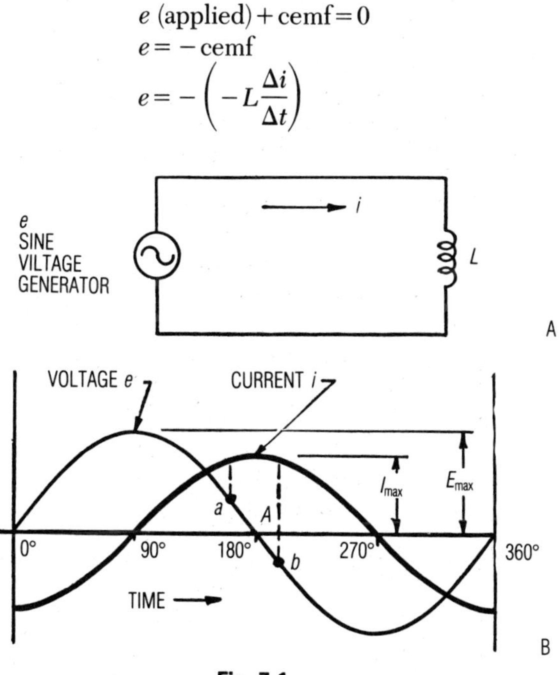

Fig. 7-1

Therefore:

$$e = +L\frac{\Delta i}{\Delta t}$$

This expression reveals that $\frac{\Delta i}{\Delta t}$, the rate of change of current, is positive when the applied voltage is positive. Thus, when $\frac{\Delta i}{\Delta t}$ is positive, or greater than zero, e is positive. When $\frac{\Delta i}{\Delta t}$ is negative, or less than zero, e is negative. And when $\frac{\Delta i}{\Delta t}$ is equal to zero, e is equal to zero.

B of Fig. 7-1 shows the voltage and current sine curves for this circuit. At a time corresponding to point a, the voltage e is positive, $\frac{\Delta i}{\Delta t}$ is positive, and therefore the current is increasing with time. At point b the voltage is negative, $\frac{\Delta i}{\Delta t}$ is less than zero, and current is decreasing with time. For some value of time, then, between points a and b, $\frac{\Delta i}{\Delta t}$ is equal to zero. That is, in going from a positive to a negative value, $\frac{\Delta i}{\Delta t}$ must go through zero. That means at point A (180° on the time axis), the curve for current must flatten out (neither increasing nor decreasing) as it passes through its maximum value. In like manner, the rest of the current curve in B may be drawn. It may be seen, then, that current is zero when the applied voltage is maximum, and minimum when the applied voltage is zero. That means the current is lagging the applied voltage by 90°.

A closer look at this phase shift across an inductance shows the relationships among the induced voltage (or cemf), the current, and the applied voltage. From Lenz's law of the generation, magnitude, and direction of the induced voltage, it is known that the cemf is a voltage opposite in phase to the applied voltage, or 180° out of phase with the applied voltage. In Fig. 7-2A this condition is shown graphically. When the applied voltage is zero, there is no opposition, and when the applied voltage is maximum, there is a maximum opposition, an induced voltage in the opposite direction.

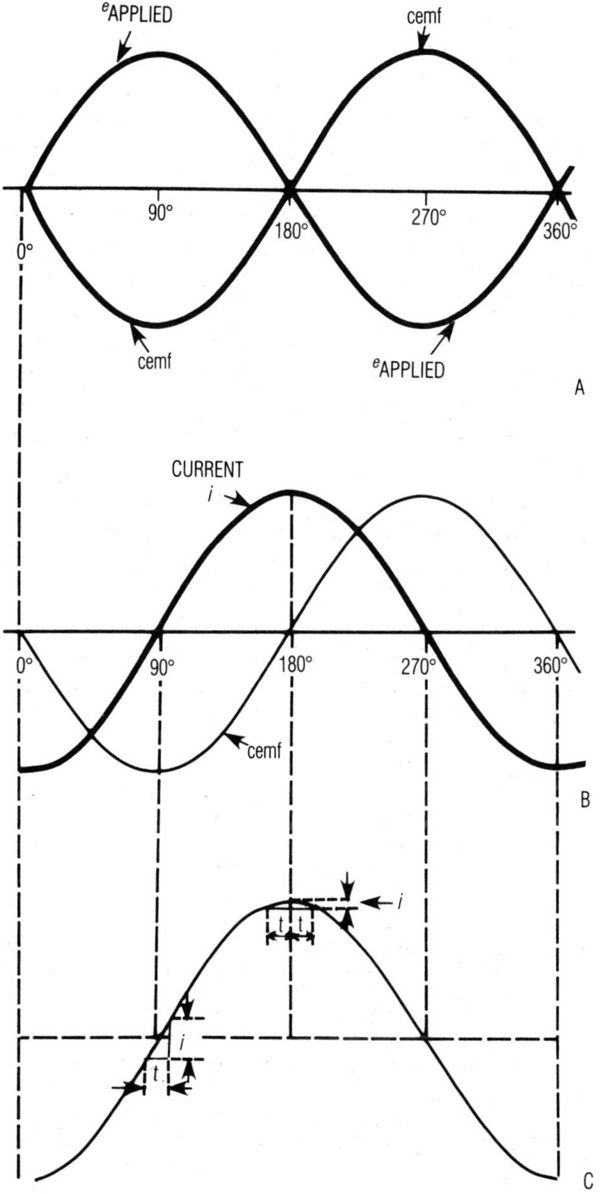

eAPPLIED
cemf
90°
180°
270°
360°
0°
cemf
eAPPLIED
A
CURRENT
i
0°
90°
180°
270°
360°
cemf
B
i
t
t
i
t
C

Fig. 7-2

A closer look at the sine wave of current through the inductance shows its phase relationship to the induced voltage. In Fig. 7-2B, it is seen that the rate of change of the current is greatest (either negative or positive) as the current passes through zero — that is, the slope of the curve at this point is steepest. This means that if a small time interval were isolated at this point, as in C, it would show a relatively large amount of current change. The induced voltage, or cemf, is greatest at the point at which the current passes through zero. In like manner, a further examination of the curve of the current in B will show that the rate of change of current is least at the point at which the current reaches a maximum value. That is, the slope of the curve at that point is least steep. A small time interval isolated at that point shows a relatively small amount of current change, or no current change, as in C. That means the counter emf is at its least, or zero. The same conditions exist for the other half of the hertz.

The sine wave of induced voltage, then, is maximum when current through the inductor is zero, and zero when the current is maximum. Also, as the current passes through zero moving in a positive direction, the slope of its curve is positive, and therefore the induced voltage is a negative maximum, since the induced voltage is always in a direction such as to oppose the action producing it (the change in current). From the point of view of phase, then, the current is said to *lead* the induced voltage by 90°. However, since the induced voltage is 180° out of phase with the applied voltage, the current goes through zero moving in either a negative or positive direction one-quarter hertz, or 90°, *after* the applied voltage. The current then is said to *lag* the applied voltage by 90°. Figure 7-3 shows these three relationships in graph form.

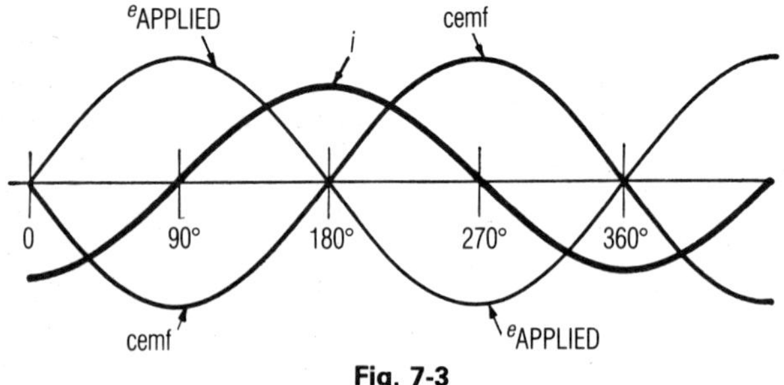

Fig. 7-3

CURRENT AND VOLTAGE ACROSS A CAPACITOR

Capacitive reactance not only increases the current flowing in an a.c. circuit, but also tends to retard the building up and the falling off of voltage. In A of Fig. 7-4, a sine wave of voltage is applied to a pure capacitance. The variation in charge on the plates of the capacitor also follows the form of the sine wave and is in phase with the voltage, since for any given capacitor the voltage across the capacitor depends directly on the charge (Fig. 7-4B and C). The current in the circuit also follows the form of the sine wave, but since a capacitor tends to retard voltage change, it is necessary to

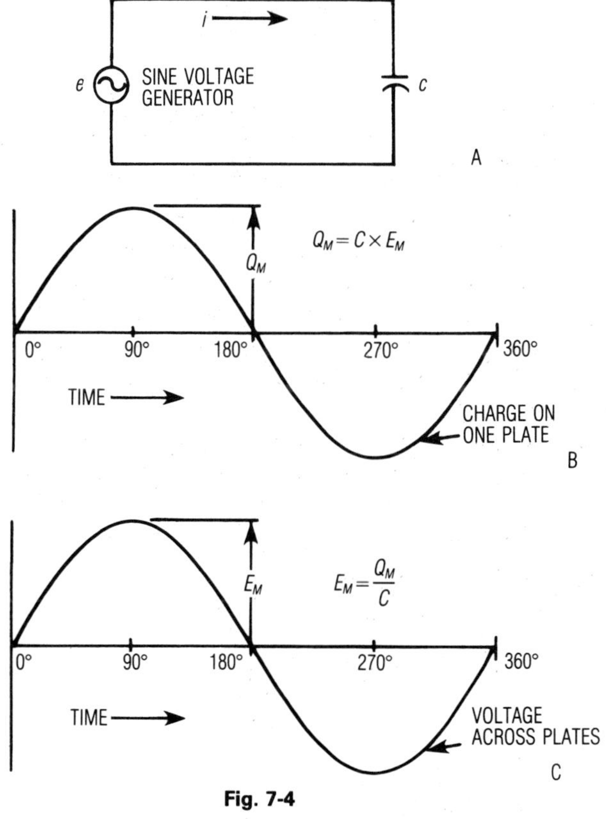

Fig. 7-4

determine precisely the delay in time between the current flow and the building up of voltage, that is, the phase shift between voltage and current.

By definition, current in any circuit is the time rate of change of charge. That means:

$$I = \frac{Q}{t}$$

And current at any instant may be expressed as:

$$i = \frac{\Delta q}{\Delta t}$$

Δq = the change in charge in coulombs,
Δt = the change in time (in seconds) required for that change,
i = the average current during that time

This expression reveals that $\frac{\Delta q}{\Delta t}$, the rate of change of charge, is positive when the current is positive. Thus, when $\frac{\Delta q}{\Delta t}$ is positive, or greater than zero, i is positive; when $\frac{\Delta q}{\Delta t}$ is negative, or less than zero, i is negative; and when $\frac{\Delta q}{\Delta t}$ is zero, i is equal to zero.

Figure 7-5A shows the sine curves of current and variations in charge for the circuit in Fig. 7-4. At a time corresponding to point a, the current i is positive, $\frac{\Delta q}{\Delta t}$ is positive, and therefore the charge is increasing with time. At point b the current is negative, $\frac{\Delta q}{\Delta t}$ is negative, and the charge is decreasing with time. For some value of time, then, between points a and b, $\frac{\Delta q}{\Delta t}$ is equal to zero; that is, in going from the positive to a negative value, $\frac{\Delta q}{\Delta t}$ must go through zero. Thus, at point A (180° on the time axis), the curve for charge must flatten out (neither increasing nor decreasing) as it passes through its maximum value. In like manner, the rest of the curve for charge is zero when the current in the circuit is maximum, and

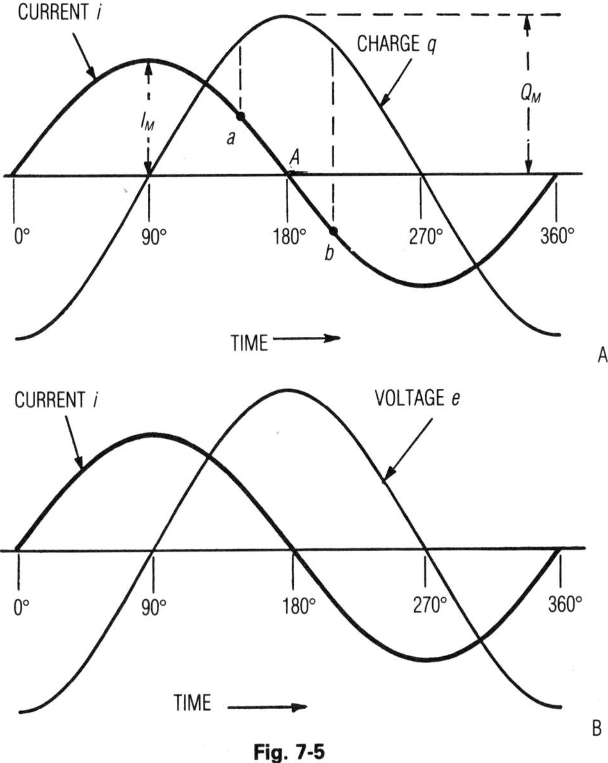

Fig. 7-5

maximum when current is zero. Therefore, the charge on the plate of a capacitor is said to lag the current through it by 90°. Since, however, the building up and falling off of charge is the building up and falling off of voltage, the voltage across the capacitor is said to lag the current through it by 90°, or the current is said to lead the voltage by 90°. See B of Fig. 7-5.

A further investigation of this phase shift across a capacitance will show the relationships among the applied voltages, the voltage drop or counter emf across the capacitor, and the current. From Kirchhoff's law it is known that the algebraic sum of the voltages around any closed circuit is zero. Therefore, the voltage across the capacitor is, by definition, a voltage opposite to the applied voltage, or 180° out of phase with the applied voltage. Figure 7-6A shows this condition graphically. Thus, when the applied voltage is zero, there is no opposition, and when the applied voltage is maximum, there is a maximum opposition, a voltage produced by the charge on the capacitor.

The voltage across the capacitor produces the current through it, and an examination of the sine wave of voltage across the capacitor

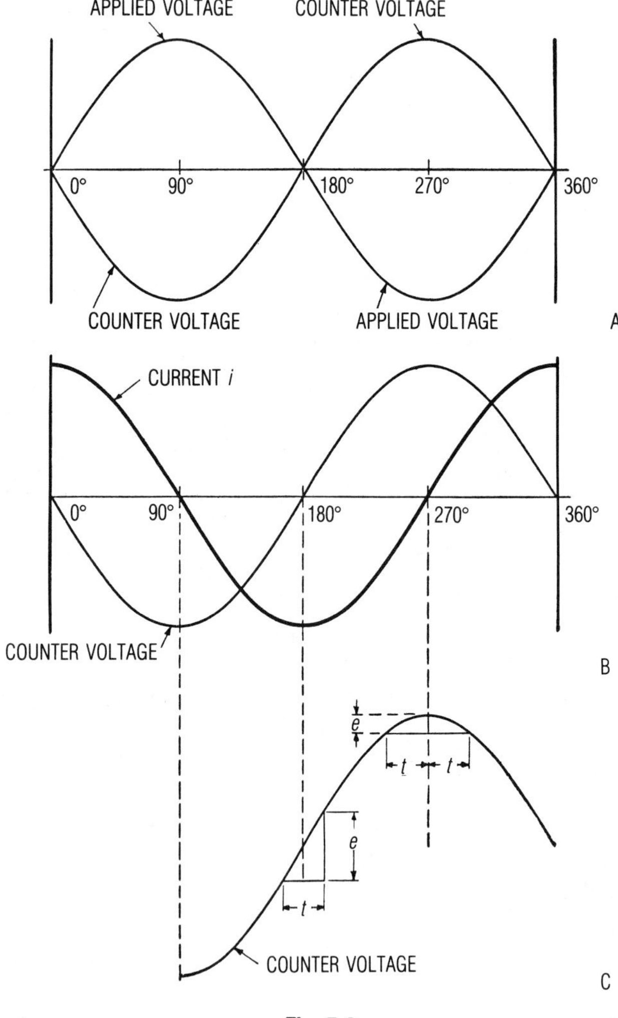

Fig. 7-6

will show its phase relationship to the current produced. In Fig. 7-6B, it will be seen that the rate of change of voltage is greatest (either negative or positive) as the voltage passes through zero (for instance, at 180°); that is, the slope of the curve at this point is steepest. This means that if a small time interval were isolated at this point, as at C, it would show a relatively large amount of voltage change. Therefore, the current is greatest at the point at which the voltage passes through zero. In like manner, a further examination of the curve of counter voltage will show that the rate of change is least at the point at which the current reaches zero value; that is, the slope of the curve is, at that point, least steep. A small time interval isolated at that point would show a relatively small amount of change in voltage, or no change in voltage, as in C. Therefore, the current would be least, or zero. The same conditions exist for the other half-hertz.

The sine wave of current, then, is at maximum when the voltage across the capacitor is zero, and zero when the voltage is maximum (Fig. 7-7). Moreover, as the counter voltage passes through zero moving in a positive direction, the slope of its curve is positive, and therefore the current is a negative maximum, since by Lenz's law the current is in a direction such as to oppose the action producing it (the change in voltage). From the point of view of phase, then, the counter voltage is said to lead the current by 90°. However, since the voltage across the capacitor is, by definition, 180° out of

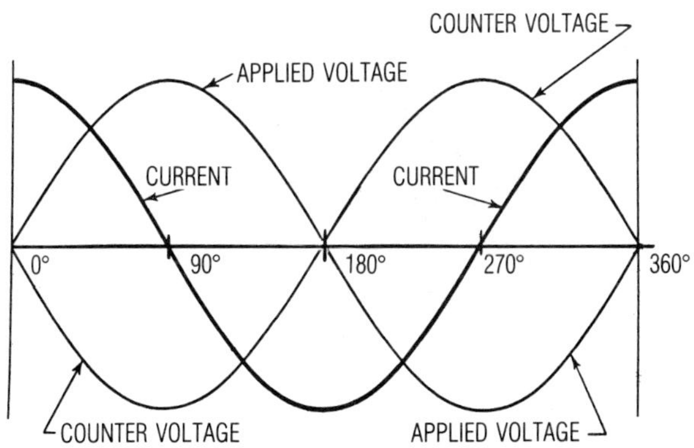

Fig. 7-7

phase with the applied voltage, the applied voltage goes through zero, moving in either a positive or negative direction 90° *after* the current. The current in a capacitor then *leads* the applied voltage by 90°.

CHAPTER 8

AC Circuits with Resistance, Inductance, and Capacitance

SERIES RL CIRCUITS

In any resistive circuit without inductance (L equal to zero), the voltage and current are said to be *in phase*. But it should be noted that in no circuit is the value of the inductance ever actually zero, since, by definition, it would then be possible to change the value of the current instantaneously. However, for all practical purposes, those circuits that do not contain inductors, or appreciable amounts of inductance, can be considered as pure resistive circuits. The time lag of the current in such a circuit is so small as to be negligible. Figure 8-1 shows this circuit and illustrates graphically the in-phase relationship of a sine wave of voltage and current across a resistance. From the graph it may be seen that the current and the voltage are alternating in polarity at the same frequency. Current rises as the applied voltage rises and is maximum. Current falls off as the voltage falls and is zero when the voltage is zero. The magnitude of the current may be determined by Ohm's law for maximum, effective, or average value, since the frequency of the current change has no effect on the in-phase relationship. Thus:

$$I_{max} = \frac{E_{max}}{R}$$

$$I_{rms} = \frac{E_{max}}{R}$$

$$I_{av} = \frac{E_{av}}{R}$$

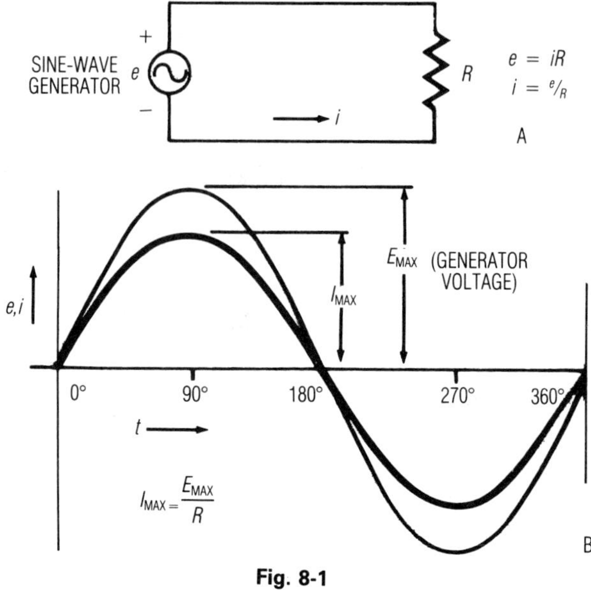

Fig. 8-1

In any circuit containing both inductance and resistance, there is a 90° phase shift of voltage and current across the inductance alone, and no phase shift across the resistance. But it must be emphasized that in any series circuit, current is the same in all parts of the circuit. Since current is the line of reference for both the inductance and resistance, it may be seen that the voltage developed across the resistance (by the current through it) will be 90° out of phase with the voltage across the inductance. Figure 8-2A shows an RL circuit. B of Fig. 8-2 shows the relationships among the current, the voltage across the inductance, and the voltage across the resistance. Thus, it may be seen that the presence of resistance in an inductive circuit results in two separate voltage drops 90° out of phase with each other. The resultant voltage of these two voltages is the voltage drop in the whole circuit and is, by Kirchhoff's law, equal to the applied voltage. The amount of phase shift of the current in such a circuit is measured not with relation to the voltage across the inductance alone (always 90°), but with relation to the resultant voltage, which is the applied voltage, E_A.

The resultant voltage and the phase angle (usually given the

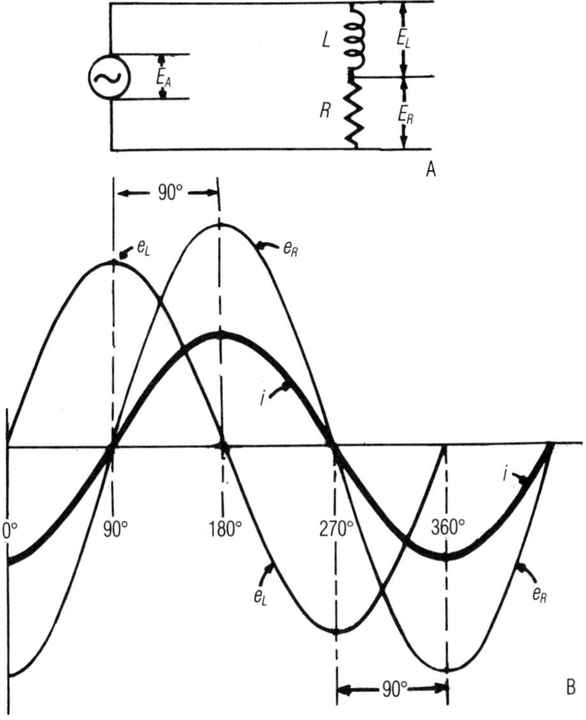

Fig. 8-2

symbol theta, θ) of any RL circuit may be determined by means of vectors. In A of Fig. 8-3, the voltage across the resistance is laid off on the horizontal vector and the voltage across the inductance on the vertical vector. Since these two voltages are 90° out of phase, the angle between them is a right angle. By drawing in a parallelogram based on these two sides, the resultant vector E_A is the hypotenuse of a right triangle. Then, by the theorem of Pythagoras, the square on the hypotenuse is equal to the sum of the squares of the other two sides, or:

$$E_A^2 = E_R^2 + E_L^2$$
$$E_A^2 = \sqrt{E_R^2 + E_L^2}$$

As stated previously, it is known that the current in the circuit is in phase with the voltage across the resistance. Therefore, the

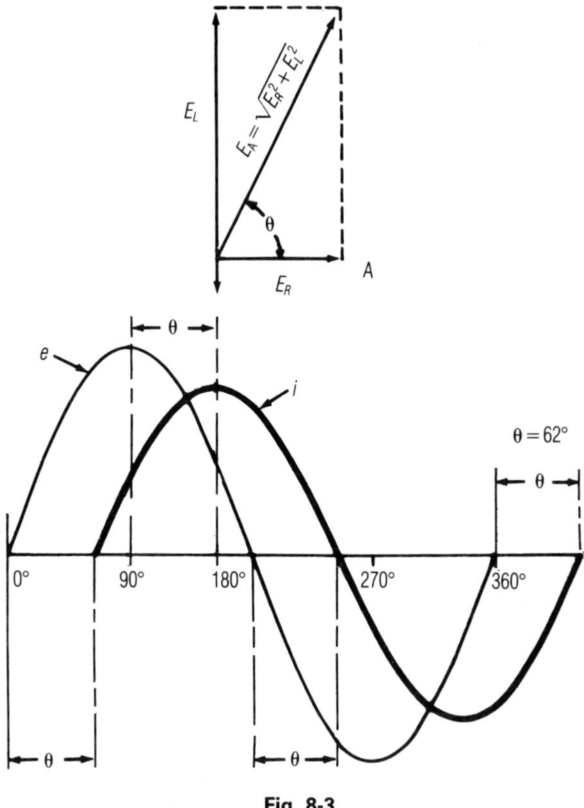

Fig. 8-3

position of the current with respect to the applied voltage is the same as the vector E_R, the voltage across the resistance. The phase angle θ then is the angle that applied voltage vector E_A makes with the vector E_R, as shown in Fig. 8-3A. The angle θ may be measured in terms of any of its trigonometric functions, depending on the values known. If the voltage across the resistance is large with respect to that across the inductance, the resultant vector approaches the horizontal and the phase angle is small. In like manner, if the voltage across the resistance is small, the resultant vector approaches the vertical and the phase angle approaches 90°. Hence, the presence of resistance in an inductive circuit causes current to lag the applied voltage by some angle *less* than 90°. B of Fig. 8-3 shows in graph form the relative positions of voltage and current and the phase angle θ.

Impedance

In any circuit containing both inductance and resistance, the total opposition offered by the circuit is *not* the simple arithmetical sum of the inductive reactances and the resistance. The inductive reactance must be added to the resistance in such a manner as to take into consideration the 90° phase difference between the two voltages in the circuit. This total opposition is termed *impedance*. Impedance is given the symbol Z. Since the voltage across the inductance is determined by the inductive reactance and the current through the inductance:

$$E_L = IX_L$$

The voltage across the resistance is determined by the resistance and the current through it:

$$E_R = IR$$

Then the resultant voltage of these two, or the applied voltage, is determined by the current and the total opposition of the circuit:

$$E_A = IZ$$

But, as previously shown:

$$E_A = \sqrt{E_R^2 + E_L^2}$$

Or:

$$E_A = \sqrt{(IR)^2 + (IX_L)^2}$$

Then:

$$IZ = \sqrt{I^2(R^2 + X_L^2)}$$
$$IZ = I\sqrt{R^2 + X_L^2}$$
$$Z = \sqrt{R^2 + X_L^2}$$

Thus, *the impedance of an RL circuit is equal to the square root of the sum of the squares of the resistance and the inductive reactance.*

The same result may be obtained more readily by means of vectors. The voltage across the resistance E_R is equal to IR, and the voltage across the inductance E_L to IX_L. Since each vector represents a product of which current is a common factor, the vectors may be laid off proportional to R and X_L and separated by 90°. Figure 8-4 shows these vectors. The resultant vector Z is the hypotenuse of a right triangle and represents the impedance of the circuit. Then:

$$Z = \sqrt{R^2 + X_L{}^2}$$

Now you can see that the angle θ is the phase angle because the direction of the impedance vector is actually the same as that of the applied voltage vector. This angle is generally determined in terms of its tangent, $\frac{X_L}{R}$, or in terms of its cosine, $\frac{R}{Z}$. Cosine is usually preferred since it translates directly into the power factor.

From the vector diagram of Fig. 8-4 you can see that if the resistance is large with relation to the inductive reactance, the circuit tends to act as a pure resistive circuit, the phase angle approaches 0°, and the impedance approaches the resistance. If the inductive reactance is large with relation to the resistance, the circuit acts as a pure inductive circuit, the phase angle approaches 90°, and the impedance approaches X_L. For practical purposes, then, the impedance of the circuit may be taken to be substantially equal to the larger quantity if the ratio of the reactance to the resistance, or of the resistance to the reactance, is 10 to 1 or greater.

The current in an a.c. circuit containing inductance and resistance may be determined by substituting the impedance Z for the resistance R used in the formula applied to d.c. circuits. Thus:

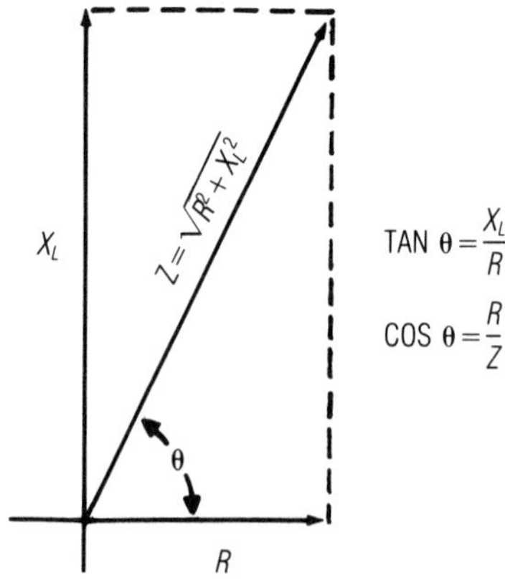

TAN $\theta = \frac{X_L}{R}$

COS $\theta = \frac{R}{Z}$

Fig. 8-4

$$I_T = \frac{E_A}{Z}$$

$$E_A = I_T Z$$

And:

$$Z = \frac{E_A}{I_T}$$

These formulas comprise Ohm's law for a.c. circuits.

Example 1

A 120-volt, 60-Hz a.c. line is connected across an inductance of 5 H and a resistance of 1,000 ohms in series. What is the inductive reactance, impedance, and the current, plus phase angle?

1. Decide what the known and unknown quantities are. Known:

   $$E_A = 120 \text{ volts} \quad F = 60 \text{ Hz}$$
   $$L = 5 \text{ H} \quad R = 1{,}000 \text{ ohms}$$

   Unknown:

   $$X_L = ? \quad Z = ? \quad I_T = ? \quad \text{angle } \theta = ?$$

2. Decide which is needed first in order to find the other unknowns. In this case the X_L is used to find the impedance, so use the formula for finding X_L.
3. The formula for X_L is $2\pi FL$.
4. Substitute in the formula to obtain:

   $$X_L = 6.28 \times 60 \times 5$$
   $$X_L = 1{,}884 \text{ ohms}$$

5. Next, find the impedance since you now have R and X_L.
6. The formula for finding impedance is:

   $$Z = \sqrt{X_L{}^2 + R^2}$$

7. Substitute in the formula to obtain:

   $$Z = \sqrt{1{,}884^2 + 1{,}000^2}$$
   $$Z = \sqrt{3{,}549{,}456 + 1{,}000{,}000}$$
   $$Z = \sqrt{4{,}549{,}456}$$
   $$Z = 2{,}133 \text{ ohms}$$

8. Now that you know the impedance and the inductive reactance, you can find the total current in the circuit. Use the Ohm's law formula for a.c. circuits.
9. The formula for finding total current is:

$$I_T = \frac{E_A}{Z}$$

10. Substitute in the formula to obtain:

$$I_T = \frac{120}{2133}$$

$$I_T = 0.056 \text{ ampere} = \text{the total current}$$

11. Now that you know the total current, the total opposition, and the total voltage, you can find the phase angle too.
12. Use the formula and substitute to obtain:

$$\text{Cos } \angle\theta = \frac{R}{Z}$$

13. Substitute in the formula to obtain:

$$\text{Cos } \angle\theta = \frac{1000}{2133} = 0.4688$$

14. Convert this to degrees by using the calculator or trig tables to obtain: $\angle\theta = 62°$.

Alternate method of finding the cosine $\angle\theta$. You can use the voltage drop across the resistor and the applied voltage to obtain the cosine, or Cos $\angle\theta = \frac{E_R}{E_A}$.

PROBLEMS

Find the missing values in the following:

	E_A (volts)	I_T (amperes)	E_L (volts)	E_R (volts)	Z (ohms)	R (ohms)	X_L (ohms)	Phase Angle (degrees)
a.	141.42	1				100	100	45
b.		2				500	100	

	E_A (volts)	I_T (amperes)	E_L (volts)	E_R (volts)	Z (ohms)	R (ohms)	X_L (ohms)	Phase Angle (degrees)
c.		3				1500	500	
d.		4				1000	200	
e.		5				10	20	
f.		6				20	10	
g.		7				50	20	
h.		8				30	20	
i.		9				40	60	
j.		10				15	30	

PARALLEL RL CIRCUITS

A of Fig. 8-5 shows an inductance L and a resistance R connected in parallel across an a.c. source. In a circuit of this type, Kirchhoff's law for parallel circuits states that the voltage across the inductance is equal to the voltage across the resistance, and that this voltage is the same as the applied voltage. That means all voltages in this circuit, being the same voltage, are in phase with each other. Keep in mind, though, that the current through the inductance lags the applied voltage by 90°. The current through the resistance is *in phase* with the applied voltage; see B of Fig. 8-5. Thus, the current in the inductance *lags* the current in the resistance by 90°. The resultant current (line current, I_T) is the vector sum of these two currents.

In Fig. 8-6, the current through the resistance, I_R, is laid off on the horizontal vector. Current through the inductance, I_L, is placed on the vertical vector. The I_L vector is laid off in the negative direction. That is because this current *lags* the current in the resistance. That vector is taken as the reference vector, since it is in phase with the applied voltage and represents also the direction of the applied voltage. Now, as in any parallel circuit, the current in the resistance is equal to the voltage divided by the resistance:

$$I_R = \frac{E_R}{R}$$

The current in the inductance is equal to the voltage divided by the inductive reactance, X_L:

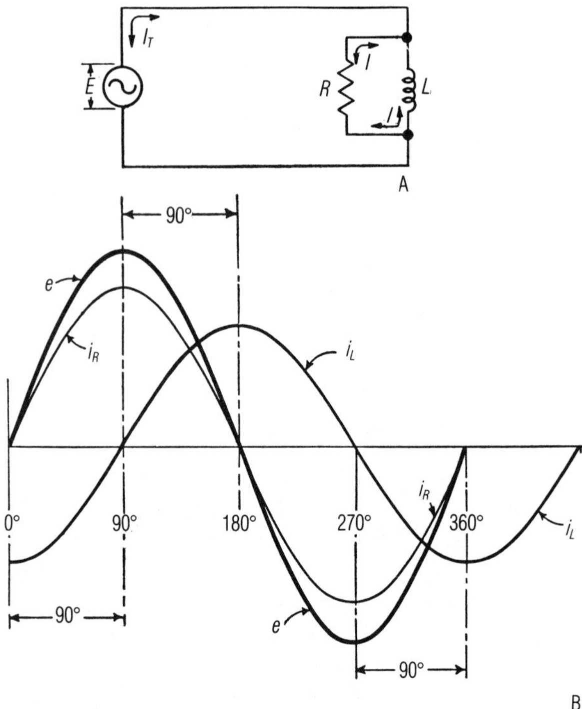

Fig. 8-5

$$I_L = \frac{E_L}{X_L}$$

The resultant vector I_T represents the total current in the circuit, and the angle this vector makes with the horizontal is the phase angle θ. That means the angle θ is the angle the line current makes with relation to the applied voltage, since the direction of the applied voltage is the same as that of the vector I_R. By convention, vectors are rotated in a counterclockwise direction. Since the I_T vector follows the E vector, the line current is said to *lag* the applied voltage by the angle θ. The tangent of this angle is then $\frac{I_L}{I_R}$. But I_L is equal to $\frac{E_L}{X_L}$ and I_R is equal to $\frac{E_R}{R}$. By substitution and cancellation:

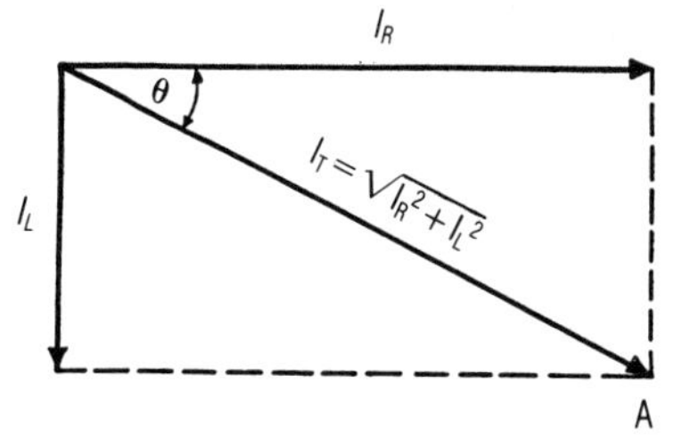

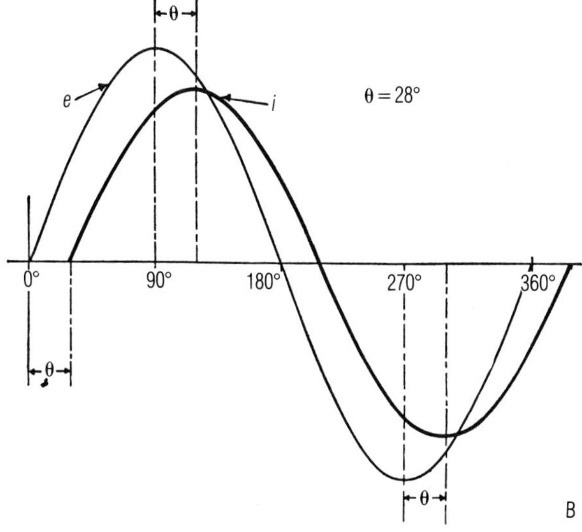

Fig. 8-6

$$\text{Tan } \angle\theta = \frac{R}{X_L}$$

$$\text{Cos } \angle\theta = \frac{I_R}{I_T} \text{ or } \frac{Z}{R}$$

The magnitude of the line current vector I_T must always be greater than either I_R or I_L because it is the hypotenuse of a right triangle. Then:

$$I_T = \sqrt{I_R^2 + I_L^2}$$

Thus, as in d.c. circuits, total current in a parallel RL circuit is always greater than the current in either branch. And the impedance of the circuit is less than the opposition of either branch. By Ohm's law:

$$Z = \frac{E_A}{I_T}$$

The impedance of a parallel RL circuit may also be obtained by the use of a formula similar to that used for resistances in parallel. Remember that the total resistance of two resistances in parallel is equal to their product divided by their sum:

$$R_T = \frac{R_1 \times R_2}{R_1 + R_2}$$

Then, by analogy:

$$Z = \frac{R \times X_L}{R + X_L}$$

But, as has been previously shown, the addition of two vector quantities $(R + X_L)$ may not be made directly. Therefore:

$$Z = \frac{RX_L}{\sqrt{R^2 + X_L^2}}$$

Example 2

What is the phase angle, total current, and impedance if the resistance in a parallel circuit is 1,000 ohms and the inductive reactance is 1,884 ohms? Applied voltage is 120, 60 Hz.

1. Identify the known and unknown. Known:

   $R = 1{,}000$ ohms
   $X_L = 1{,}884$ ohms
   $E_A = 120$ volts

   Unknown:

   Phase angle = ?

   $I_T = ?$

   $Z = ?$

2. Decide which must be found first.

3. Use the formula for finding the impedance:

$$Z=\frac{RX_L}{\sqrt{R^2+X_L{}^2}}$$

4. Substitute the values in the formula:

$$Z=\frac{1{,}000\times 1{,}884}{\sqrt{1{,}000{,}000+3{,}549{,}456}}$$

$$Z=\frac{1{,}884{,}000}{2{,}132.945381}=883.2856278 \text{ ohms}$$

5. Now that you have impedance, you can find the line current, I_T:

$$I_T=\frac{E_A}{Z}$$

6. Substitute the values into the formula to obtain:

$$I_T=\frac{120}{883.2856278}=0.1358563937 \text{ ampere}$$

7. The phase angle can now be found since you know the total opposition (Z) and the resistance.
8. Substitute the values in the formula to obtain:

$$\text{Cos}\angle\theta=\frac{I_R}{I_T} \text{ or } \frac{Z}{R}$$

$$\text{Cos }\angle\theta=\frac{883.2856278}{1{,}000}=0.8833$$

$$\angle\theta=27.95696075°$$

Problems

1. What is the phase angle when the impedance is 100 ohms and the resistance is 500 ohms in a parallel RL circuit with a 100 V a.c. power source?
2. What is the phase angle when the impedance is 1,000 ohms and the resistance is 1,500 ohms in a parallel RL circuit with a 100 V a.c. power source?

3. What is the total current in a parallel RL circuit when the applied voltage is 100 volts and the impedance is 50 ohms?
4. What is the total current in a parallel RL circuit when the applied voltage is 1,000 volts and the impedance is 5,000 ohms?
5. What is the voltage across the resistor and the inductor in a parallel RL circuit when the applied voltage is 120?
6. What are the missing values in the table below?

	E_A	I_T	Z	I_R	I_L	R	X_L	Phase Angle	E_L	E_R
a.	100					50	50			
b.	200			3	5					
c.	300			4	6					
d.	400			5	10					
e.	500			6	12					
f.	600			7	14					
g.	700			8	15					
h.	800			9	10					
i.	900			10	12					
j.	1000			11	20					

PARALLEL RC CIRCUITS

Figure 8-7 shows capacitance C and resistance R connected in parallel across an a.c. source. Since this is a parallel circuit, voltage is the same across all components. All voltages are therefore in phase with each other. However, the current through the capacitor (c) leads the applied voltage by 90°. The current through the resistance is in phase with the applied voltage, as in B. Thus, the capacitive current leads the resistive current by 90°, and the resultant current, or line current, is the vector sum of these two currents. A of Fig. 8-8 shows the current through the resistance I_R is laid off on the horizontal vector, and the current through the capacitance I_C on the vertical vector. The I_C vector is laid off in the positive direction because this current leads the resistive current, which is taken as the reference vector, since it is in phase with the applied voltage and represents the direction of the applied voltage. The resultant

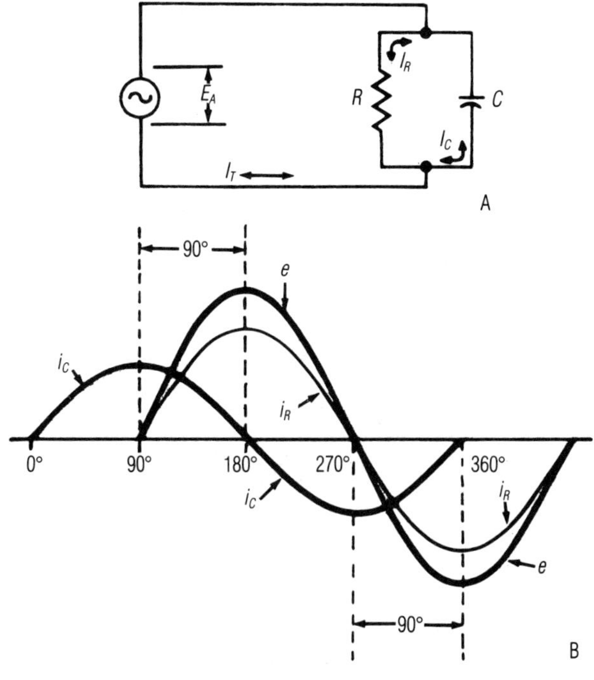

Fig. 8-7

vector I_T represents the total current in the circuit, and the angle this vector makes with the horizontal is the phase angle θ.

The line current, then, is said to lead the applied voltage by the angle θ, since by conventional rotation the I_R vector follows the I_T vector (see B of Fig. 8-8). The tangent of this angle is $\frac{I_C}{I_R}$, but as in any parallel circuit:

$$I_c = \frac{E}{X_C}$$

$$I_R = \frac{E}{R}$$

And:

$$I_T = \frac{E}{Z}$$

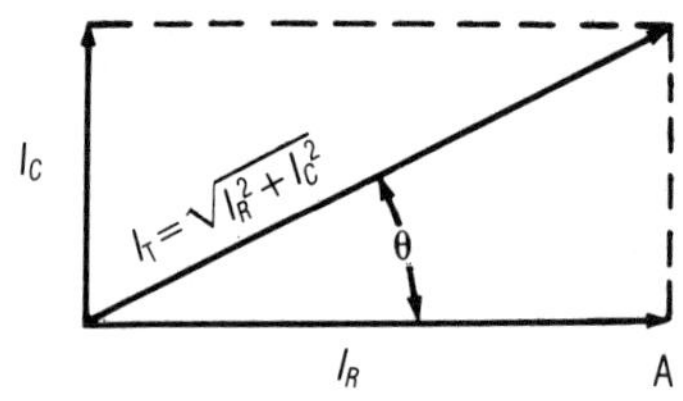

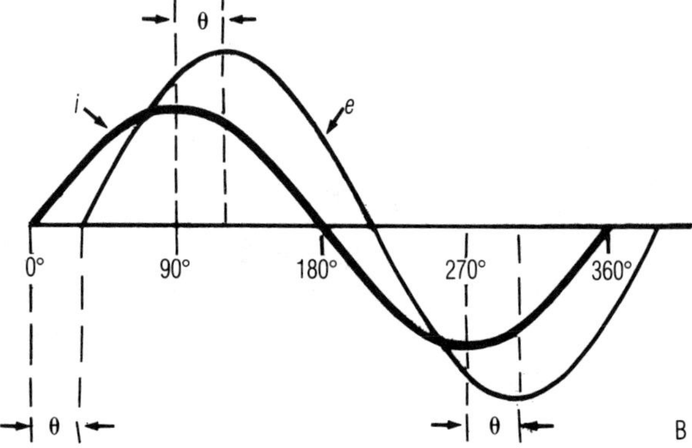

Fig. 8-8

Then, by substitution and cancellation:

$$\text{Tan } \angle\theta = \frac{R}{X_C}$$

$$\text{Cos } \angle\theta = \frac{Z}{R}$$

The magnitude of the line current vector I_T must always be greater than either I_R or I_C because it is the hypotenuse of a right triangle. Then:

$$I_T = \sqrt{I_R^2 + I_C^2}$$

Thus, it may be seen that, as in d.c. circuits, total current in a parallel RC circuit is always greater than the current in either branch, and, by extension, the impedance of the circuit is less than the opposition of either branch.

Impedance

The impedance of a parallel RC circuit may also be obtained by the use of a formula similar to that used for resistances in parallel. Since:

$$R_T = \frac{R_1 \times R_2}{R_1 + R_2}$$

Then:

$$Z = \frac{R \times X_C}{R + X_C}$$

But, as has been previously shown, the addition of two vector quantities $(R + X_C)$ may not be made directly. Thus:

$$Z = \frac{RX_C}{\sqrt{R^2 + X_C{}^2}}$$

For example, if the values used previously for the series RC circuit are transferred to the parallel RC circuit in A of Fig. 8-7, then, as before:

$R = 1{,}000$ ohms
$X_C = 2{,}652.582384$ ohms

Then:

$$Z = \frac{RX_C}{\sqrt{R^2 + X_C{}^2}} = \frac{1{,}000 \times 2{,}652.582384}{\sqrt{1{,}000{,}000 + 7{,}036{,}193.318}}$$

$$Z = \frac{2{,}652{,}582.384}{2{,}834.81804}$$

$$Z = 935.7152193 \text{ ohms}$$

The line current is:

$$I_T = \frac{120\text{V}}{935.7152193\Omega}$$

$$I_T = 0.1282441469 \text{ ampere}$$

The line current leads the applied voltage by the angle θ:

$$\text{Cos } \angle\theta = \frac{I_R}{I_T}$$

Since we don't have I_R, it has to be found:

$$I_R = \frac{E_R}{R} = \frac{120\text{V}}{1{,}000\Omega} = 0.120 \text{ ampere}$$

Then we can go back to finding the phase angle:

$$\text{Cos } \angle\theta = \frac{0.120}{0.1282441469}$$
$$\text{Cos } \angle\theta = 0.935715219$$
$$\angle\theta = 20.65599758°$$

If the values of frequency, capacitance, and resistance are the same, the angles in series and parallel are complementary. Or, if you add the two angles, the result is 90°. So the parallel angle is 20.65599758° and the series angle is 69.34400242°. Add the two and you get 90°.

Example 3

A parallel circuit consisting of a capacitor of 100 μF and a resistance of 50 ohms has an applied voltage of 120 at 60 Hz. What is the total current, the impedance, the individual circuit component currents, and the phase angle?

1. Decide what has to be found and establish an order in which they may be found. Some depend on others, so you have to do one before the other in some cases.

   We know that the voltage across the capacitor and resistor are the same. The voltage is that which is applied. In this case the voltages across the capacitor and the resistor are 120 volts. In order to find the currents you have the resistance, but not the X_C. So, find I_C when you have X_C. That means X_C is probably the first thing to find.

2. Find $X_C = \frac{1}{2\pi fC}$.

3. Substitute the values into the formula to obtain:

$$X_C = \frac{1}{6.283185308 \times 60 \times 100 \times 10^{-6}}$$
$$X_C = 26.52582384 \text{ ohms}$$

4. Now you can find the individual currents.

$$I_C = \frac{120V}{26.52582384\Omega}$$
$$I_C = 4.523893423 \text{ amperes}$$

Next:

$$I_R = \frac{120V}{50\Omega}$$
$$I_R = 2.4 \text{ amperes}$$

5. Now you can find the total current by using the formula:

$$I_T = \sqrt{I_R^2 + I_C^2}$$

6. Substitute the currents in the formula and use the calculator to obtain an answer:

$$I_T = \sqrt{(2.4)^2 + (4.523893423)^2}$$
$$I_T = 5.1210947777 \text{ amperes}$$

7. Now it is possible to find the impedance by using the formula:

$$Z = \frac{E_A}{I_T}$$
$$Z = \frac{120}{5.1210947777}$$
$$Z = 23.43248958 \text{ ohms}$$

Note: The impedance is less than either the R or X_C, which is normal for a parallel circuit.

8. Now we can find the phase angle by substituting in the formula:

$$\text{Cos } \angle\theta = \frac{I_R}{I_T}$$
$$\text{Cos } \angle\theta = \frac{2.4}{5.1210947777}$$
$$\text{Cos } \angle\theta = 0.4686497916$$
$$\angle\theta = 62.05331277^\circ$$

Example 4

A parallel circuit consisting of a capacitor of 0.01 μF and a resistance of 250 ohms is connected to a signal generator producing a 10-volt signal at 100 kHz. What is the current through the capacitor and resistor, the total current, and the impedance and phase angle?

1. Decide what has to be found and establish an order in which they may be found. Some depend on others, so you have to find one before the others in some cases. We *know*: The voltage across the capacitor and resistor are the same. The voltage is that which is applied or generated by the signal source. In this case the voltages across the capacitor and the resistor are 10 volts.

   In order to find the currents you have the resistance, but not the X_C. So, find I_C when you have X_C. That means X_C is probably the first thing to find.

2. Find $X_C = \frac{1}{2\pi f C}$.

3. Substitute the values into the formula to obtain:

$$X_C = \frac{1}{6.283185308 \times 100 \times 10^3 \times 0.01 \times 10^{-6}}$$
$$X_C = 159.1549431 \text{ ohms}$$

4. Now you can find the individual currents:

$$I_C = \frac{10}{159.1549431}$$
$$I_C = 0.062831853 \text{ ampere}$$

5. Next:

$$I_R = \frac{10}{250}$$
$$I_R = 0.04 \text{ ampere}$$

6. Now you can find the total current by using the formula:

$$I_T = \sqrt{I_R{}^2 + I_C{}^2}$$

7. Substitute the currents in the formula and use the calculator to obtain an answer:

$$I_T = \sqrt{I_R^2 + I_C^2}$$
$$I_T = \sqrt{(0.04)^2 + (0.062831853)^2}$$
$$I_T = 0.0744838354 \text{ ampere}$$

8. Now it is possible to find the impedance by using the formula:

$$Z = \frac{E_A}{I_T}$$
$$Z = \frac{10}{0.0744838354}$$
$$Z = 134.2573182 \text{ ohms}$$

Note: The impedance is less than either the R or X_C, which is normal for parallel circuits.

9. Now we can find the phase angle by substituting in the formula:

$$\text{Cos } \angle\theta = \frac{I_R}{I_T}$$
$$\text{Cos } \angle\theta = \frac{0.04}{0.0744838354}$$
$$\text{Cos } \angle\theta = 0.5370292733$$
$$\angle\theta = 57.51836333°$$

PROBLEMS

1. What is the impedance of a parallel RC combination that has 0.001 μF capacitance and 250 ohms resistance with an applied voltage of 100 volts at 100 Hz?
2. What is the impedance of a parallel RC combination that has 0.01 μF and 250 ohms with an applied voltage of 100 volts and 100 Hz?
3. What is the impedance of a parallel RC combination that has a 0.01μF capacitor and 25 ohms with an applied voltage of 100 volts at 100 Hz?
4. What is the impedance of a parallel RC combination that has a 1-μF capacitor and 2,000 ohms with an applied voltage of 120 volts at 60 Hz?

5. What is the impedance of a parallel RC combination that has a 10-μF capacitor and 20,000 ohms resistance with an applied voltage of 1,100 volts at 60 Hz?
6. What is the total current for Problem 1?
7. What is the total current for Problem 2?
8. What is the total current for Problem 3?
9. What is the total current for Problem 4?
10. What is the total current for Problem 5?
11. What is the current through the resistor in Problem 1?
12. What is the current through the resistor in Problem 2?
13. What is the current through the resistor in Problem 3?
14. What is the current through the resistor in Problem 4?
15. What is the current through the resistor in Problem 5?
16. What is the capacitor current in Problem 1?
17. What is the capacitor current in Problem 2?
18. What is the capacitor current in Problem 3?
19. What is the capacitor current in Problem 4?
20. What is the capacitor current in Problem 5?
21. Find the missing values:

	E_A (V)	I_T (A)	I_R (A)	I_C (A)	f (Hz)	C (μF)	R (Ω)	Z (Ω)	X_C (Ω)
a.	100				100	0.01	100		
b.	200				200	0.1	1000		
c.	300				300	0.1	1000		
d.	50				50	10.0	10,000		
e.	10				10k	0.001	100,000		

SERIES RC CIRCUITS

The total opposition offered by a circuit containing both a reactive element and a resistance is not the simple arithmetical sum of the reactance X and the resistance R. The capacitive reactance is added to the resistance in such a manner as to take into account the 90°

phase difference between the two voltages in the circuit. Since the impedance Z, or total opposition, of an RC circuit is equal to the applied voltage divided by the current in the circuit, then:

$$E_A = IZ$$
$$E_A = \sqrt{E_R^2 + E_C^2}$$

Or:

$$E_A = \sqrt{(I^2R^2) + (I^2X_C^2)}$$

Then:

$$IZ = \sqrt{I^2(R^2 + X_C^2)}$$
$$IZ = I\sqrt{R^2 + X_C^2}$$
$$Z = \sqrt{R^2 + X_C^2}$$

Thus, the impedance of an RC circuit is equal to the square root of the sum of the squares of the resistance and the capacitive reactance.

The same result may be obtained more readily by means of vectors. The voltage across the resistance E_R is equal to IR, and the voltage across the capacitor E_C to IX_C. Since each vector represents a product of which current is a common factor, the vectors may be laid off proportional to R and X_C, which are separated by 90°. Figure 8-9 shows these vectors. The resultant vector Z is the hypotenuse of a right triangle and represents the impedance of the circuit. Then:

$$Z = \sqrt{R^2 + X_C^2}$$

It will be seen also that the angle θ is the phase angle because the direction of the impedance vector is the same as that of the applied voltage vector. This angle is generally determined by its

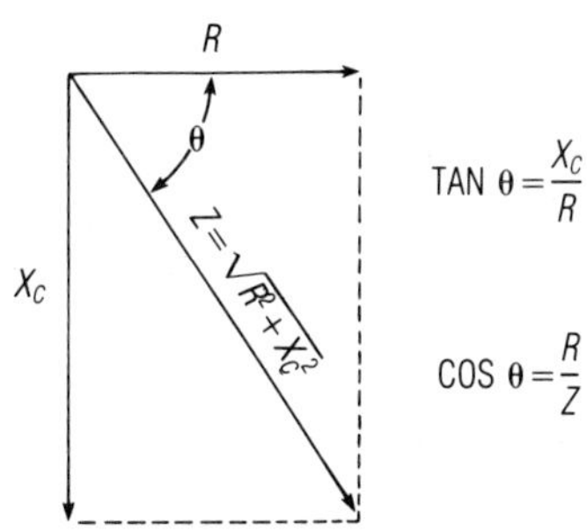

Fig. 8-9

tangent, $\frac{X_C}{R}$, or in terms of its cosine, $\frac{R}{Z}$. The cosine is usually used since it tells you the power factor also.

From the vector diagram of Fig. 8-9 it may be seen that if the resistance is large in respect to the capacitive reactance, the circuit tends to act as a pure resistive circuit, the phase angle approaches 0°, and the impedance approaches R. If the capacitive reactance is large in respect to the resistance, the circuit tends to act as a pure capacitive circuit, the phase angle approaches 90°, and the impedance approaches X_C. For practical purposes, then, the impedance of the circuit may be taken to be substantially equal to the larger value, if the ratio of the reactance to the resistance, or of the resistance to the reactance, is 10 to 1 or greater.

Example 5

What is the phase angle of a series RC circuit if 120 volts, 60 Hz are connected to a circuit of 1 μF and a resistance of 1,000 ohms? Also find the current in the circuit.

1. Find the capacitive reactance since we will need it in finding the impedance. And impedance will be needed to get $\frac{R}{Z}$ or the cosine of $\angle\theta$. Therefore:

$$X_C = \frac{1}{2\pi f C}$$

2. Plug in the values in the formula to obtain:

$$X_C = \frac{1}{6.283185308 \times 60 \times 1 \times 10^{-6}}$$

3. Solve with calculator to obtain $X_C = 2{,}652.582384$ ohms.
4. R is given as 1,000 ohms.
5. Now find the impedance (Z):

$$Z = \sqrt{R^2 + X_C{}^2}$$

6. Plug in the values to obtain:

$$Z = \sqrt{(1{,}000)^2 + (2{,}652.582384)^2}$$
$$Z = \sqrt{1{,}000{,}000 + 7{,}036{,}193.318}$$

$$Z = \sqrt{8{,}036{,}193.318}$$
$$Z = 2{,}834.81804$$

7. The current in this circuit is:

$$I = \frac{E}{Z}$$

8. Plug in the values to obtain:

$$I = \frac{120\text{V}}{2{,}834.81804\Omega}$$
$$I = 0.0423307592 \text{ ampere}$$

9. The phase angle is next to be found:

$$\text{Cos } \angle\theta = \frac{R}{Z}$$

10. Plug in the values to obtain:

$$\text{Cos } \angle\theta = \frac{1{,}000}{2{,}834.81804}$$
$$\text{Cos } \angle\theta = 0.3527563272$$
$$\angle\theta = 69.34400264°$$

Example 6

A series RC circuit has a resistance of 5,000 ohms and a capacitor with 10 microfarads of capacitance. What is the current through the circuit and what is the phase angle if the applied voltage is 100 and the frequency is 50 Hz?

1. Find the capacitive reactance since we will need it in finding the impedance. And impedance will be needed to find the current and the phase angle. Therefore:

$$X_C = \frac{1}{2\pi fC}$$

2. Plug in the values in the formula to obtain:

$$X_C = \frac{1}{6.283185308 \times 50 \times 10 \times 10^{-6}}$$

3. Solve with a calculator to obtain $X_C = 318.3098861$ ohms
4. R is given as 5,000 ohms.
5. Now find the impedance (Z):

$$Z = \sqrt{R^2 + X_C{}^2}$$

6. Plug in the values to obtain:

$$Z = \sqrt{(5{,}000)^2 + (318.3098861)^2}$$
$$Z = \sqrt{25{,}000{,}000 + 101{,}321.184}$$
$$Z = \sqrt{25{,}101{,}321.18}$$
$$Z = 5{,}010.121873$$

7. The current in this circuit is:

$$I = \frac{E}{Z}$$

8. Plug in the values to obtain:

$$I = \frac{100}{5{,}010.121873}$$
$$I = 0.0199595943 \text{ ampere}$$

9. The phase angle is to be found next:

$$\text{Cos } \angle\theta = \frac{R}{Z}$$

10. Plug in the values to obtain:

$$\text{Cos } \angle\theta = \frac{5{,}000}{5{,}010.121873}$$
$$\text{Cos } \angle\theta = 0.9979797152$$
$$\angle\theta = 3.642646851°$$

PROBLEMS

1. What is the impedance of a series RC circuit with the capacitor rated at 100 μF and the resistance at 47,000 ohms on a 220-volt, 50-Hz line?
2. What is the impedance of a series RC circuit with the

capacitor rated at 10 μF and the resistance at 470,000 ohms on a 220-volt, 50-Hz line?

3. What is the impedance of a series RC circuit with the capacitor rated at 1 μF and the resistance at 4.7 megs on a 220-volt, 50-Hz line?
4. What is the impedance of a series RC circuit with the capacitor rated at 0.1 μF and the resistance at 4.7 megs on a 220-volt, 50-Hz line?
5. What is the total current for Problem 1?
6. What is the total current for Problem 2?
7. What is the total current for Problem 3?
8. What is the total current for Problem 4?
9. What is the phase angle for Problem 1?
10. What is the phase angle for Problem 2?
11. What is the phase angle for Problem 3?
12. What is the phase angle for Problem 4?
13. What is the capacitive reactance for Problem 1?
14. What is the capacitive reactance for Problem 2?
15. What is the capacitive reactance for Problem 3?
16. What is the capacitive reactance for Problem 4?
17. From these problems, what effect do you see the capacitors having on phase angle?

CURRENT AND VOLTAGE IN AN RC CIRCUIT

In any circuit containing capacitance and resistance, there is a 90° shift of current and voltage across the capacitance and no phase shift across the resistance. Current in a series circuit is everywhere the same and is therefore taken as the *line of reference* for both the capacitance and the resistance. Since the voltage across the resistance is in phase with the current through it, and the voltage across the capacitance is 90° out of phase with the same current, it will be seen that these two voltages are 90° out of phase with each other. In Fig. 8-10, A shows a series RC circuit, and B the relationship between the current and the voltages across the elements. The

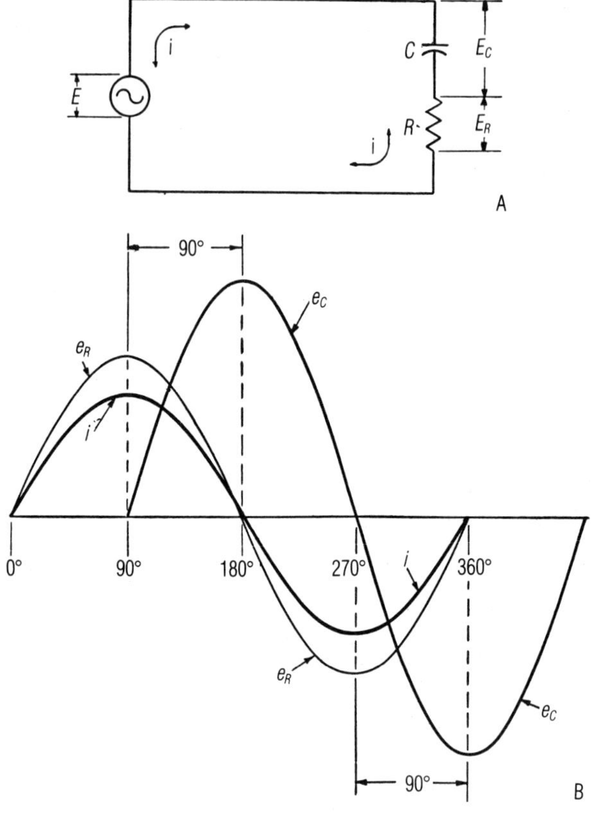

Fig. 8-10

voltage which results from the two voltage drops that are 90° out of phase is the voltage drop in the whole circuit and is, by Kirchhoff's law, equal to the applied voltage.

The relationship between the applied voltage and the voltage drops, and the phase angle, of any series RC circuit may be determined by means of vectors. In A of Fig. 8-11 the voltage across the resistance is laid off on the horizontal vector and the voltage across the capacitance on the vertical vector. Since these two voltages are 90° out of phase, the angle between them is a right triangle. By drawing in a parallelogram based on the two sides, the resultant vector E_A is seen to be the hypotenuse of a right triangle. Then,

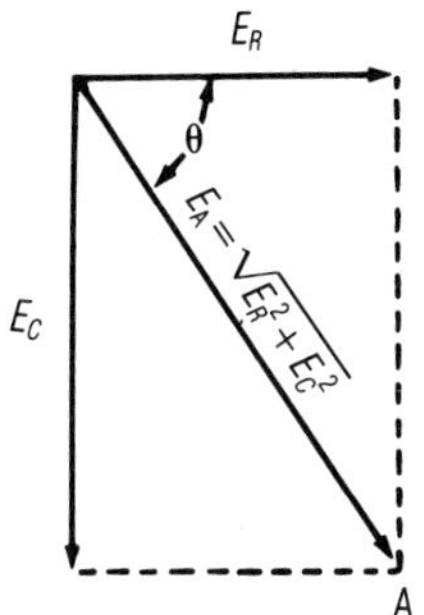

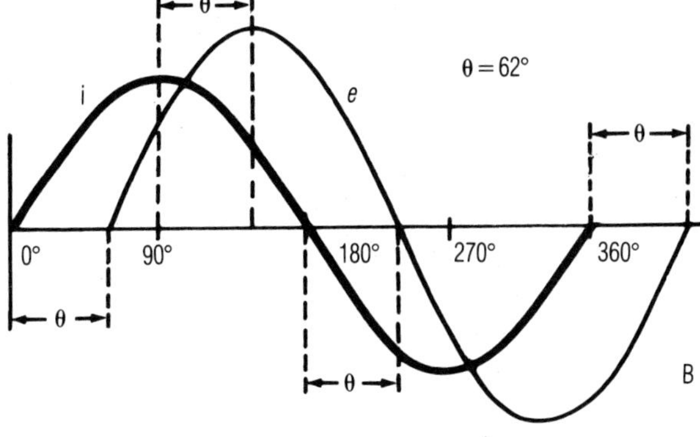

Fig. 8-11

by the theorem of Pythagoras, *the square of the hypotenuse is equal to the sum of the squares of the individual legs or two sides of the triangle.* When stated in a formula, it becomes:

$$E_A{}^2 = E_R{}^2 + E_C{}^2$$

Then, take square root of both sides of the equals sign.

$$E_A = \sqrt{E_R{}^2 + E_C{}^2}$$

Since it is known that the current in the circuit is in phase with the voltage across the resistance, the direction of the current vector is the same as the vector E_R, the voltage across the resistance. The phase angle θ then is the angle that the applied voltage E_A makes with the vector E_R, as seen in Fig. 8-11A. If the voltage across the resistance is large in respect to that across the capacitance, the resultant vector will approach the horizontal and the phase angle will be small; and, in like manner, if the voltage across the resistance is small, the resultant vector will approach the vertical and the phase angle will approach 90°. Hence, the presence of resistance in a capacitive circuit causes the current to lead the applied voltage by some angle *less* than 90°. B of Fig. 8-11 shows in graph form the relative positions of current and voltage and the phase angle θ.

In order to find the phase angle of the RC circuit with the applied voltage and the voltage across the resistor known, you use the formula:

$$\text{Cos } \angle\theta = \frac{E_R}{E_A}$$

Example 7

What is the applied voltage for a circuit with a series resistor and capacitor which have a voltage drop of 88 volts across the resistor and 90 volts across the capacitor?

1. Locate the correct formula:

$$E_A = \sqrt{E_R{}^2 + E_C{}^2}$$

   Since both the voltage drop across the resistor E_R and the voltage drop across the capacitor E_C are known, just substitute and solve.

2. Substitute the values into the formula:

$$E_A = \sqrt{(88)^2 + (90)^2}$$
$$E_A = \sqrt{7{,}744.000008 + 8{,}099.999999}$$
$$E_A = \sqrt{15{,}844}$$
$$E_A = 125.8729518 \text{ volts}$$

Example 8

What is the phase angle when the voltage across a series resistor is 100 volts and the voltage drop across the capacitor in series with the resistor reads 95 volts?

1. Find the correct formula. In this case it means you have to find the applied voltage first, since it is not given. But the

voltages for resistance and capacitance are given, so use them to obtain the applied voltage:

$$E_A = \sqrt{E_R{}^2 + E_C{}^2}$$

2. Substitute the values in the formula to obtain:

$$E_A = \sqrt{(100)^2 + (95)^2}$$
$$E_A = \sqrt{10{,}000 + 9{,}025}$$
$$E_A = \sqrt{19{,}025}$$
$$E_A = 137.9311423 \text{ volts}$$

3. Now that you have the applied voltage, and the voltage dropped across the resistor was given as 100, you can use the formula to obtain the angle.

$$\text{Cos } \angle\theta = \frac{E_R}{E_A}$$

4. Substitute into the formula the given values:

$$\text{Cos } \angle\theta = \frac{100}{137.9311423} = 0.7249994333$$

5. To convert this to degrees, use the calculator or look it up in the trig tables. It becomes:

$$\text{Cos } \angle\theta = 43.53119931°$$

PROBLEMS

1. Find the applied voltage and phase angle for the following:

	E_A (volts)	E_R (volts)	E_C (volts)	$\angle\theta$ (degrees)
a.		100	100	
b.		50	100	
c.		25	50	
d.		5	10	
e.		1,000	500	
f.		500	100	
g.		1,500	500	

	E_A (volts)	E_R (volts)	E_C (volts)	$\angle\theta$ (degrees)
h.		2,000	1,500	
i.		20	100	
j.		5	100	

2. Find the phase angle of the following voltages and find the missing voltage:

	Phase Angle (degrees)	E_A (volts)	E_R (volts)	E_C (volts)
a.	45		100	100
b.	60		50	86.6025404
c.			100	1,000
d.			50	25
e.			5	10
f.			20	30
g.			10	15
h.			150	300
i.			250	500
j.	30		86.6025404	50

CHAPTER 9

Resonance in Circuits

RESONANCE

The resonant circuit offers to electricity or tuned circuits advantages which may be utilized for purposes of discrimination or the separation of voltages at one frequency from those at another frequency. Tuned circuits may offer high voltages at certain desirable frequencies and low voltages at all other frequencies, or they may be used to short-circuit undesirable frequencies and allow all others to pass at high voltages. Thus, a certain electrical advantage may be gained from tuned circuits.

The condition of resonance is characterized by the following facts:

1. The inductive and capacitive reactances, being equal andopposite in direction, *cancel* each other. That means:

$$X_L = X_C$$

2. The frequency of the circuit at resonance, f_r, is determined by solving the formula for f.

$$f_r = \frac{1}{2\pi\sqrt{LC}}$$

3. In a *series resonant* circuit, *impedance* is at a *minimum* and *current* at a *maximum*, since the only effective opposition in the circuit is the resistance R. That means in a series RCL circuit:

$$Z = \sqrt{R^2 + (X_L - X_C)^2}$$

At resonance $Z = R$ because $X_L - X_C = 0$.

4. In a *parallel resonant* circuit, *impedance in the line* is at a *maximum* and *current in the line* is at a *minimum*. This is because the reactances act together in such a manner as to raise the impedance. That means the total impedance:

$$Z_T = \frac{X_L X_C}{X_L + X_C}$$

Then, as the sum of X_L and X_C approaches zero, the impedance in the line approaches a maximum.

Frequency at Resonance

The condition of resonance in any circuit is determined initially by the equal and opposite reactances in the circuit. Since both inductive reactance and capacitive reactance depend directly on the frequency of the applied voltage, the condition of resonance may be stated as:

$$X_L = X_C$$

That means:

$$X_L = 2\pi f L$$

And:

$$X_C = \frac{1}{2\pi f C}$$

So:

$$2\pi f L = \frac{1}{2\pi f C}$$

Then:

$$4\pi^2 f^2 LC = 1$$

And:

$$f^2 = \frac{1}{4\pi^2 LC}$$

Taking the square root of each side leaves:

$$f = \frac{1}{2\pi\sqrt{LC}}$$

Since f is the frequency when X_L is equal to X_C, it is the frequency at resonance and may be written as f_r. Then:

$$f_r = \frac{1}{2\pi\sqrt{LC}}$$

L is the inductance of the circuit in henrys
C is the capacitance of the circuit in farads
f_r is the resonant frequency of the circuit

Now let's take an example and use it to see if the formula can find a resonant frequency for a combination of components.

Example 1

What is the resonant frequency of a circuit with the following components: L is 2 mH, C is 80 pF?

1. Determine the components needed and find a formula to fit the components to obtain the resonant frequency.
2. State the formula:

$$f_r = \frac{1}{2\pi\sqrt{LC}}$$

3. Substitute the values in the formula:

$$f_r = \frac{1}{6.28 \times \sqrt{0.002 \times 0.00000000008}}$$

or, if your calculator cannot display all the zeros, use the powers of 10 and then your exponents on the calculator.

$$f_r = \frac{1}{6.28 \times \sqrt{2 \times 10^{-3} \times 8 \times 10^{-11}}}$$

4. Solve the problem with your calculator to obtain:

$$f_r = \frac{1}{2.512 \times 10^{-6}}$$

$f_r = 398{,}089.172$ hertz or 398 kHz

Problems

1. Fill in the missing values in the table below:

	f_r (Hz) Frequency	L (H) Inductance	C (μF) Capacitance
a.		1	7
b.		10	7
c.		0.1	7
d.		0.001	7
e.		1	14
f.		10	14
g.		0.1	0.1
h.		0.001	0.1

2. What happened to the resonant frequency as the inductance was increased and the capacitance stayed the same?
3. What happened to the resonant frequency as the inductance was held the same (say, at 1 H) while the capacitance was increased from 7 to 14 μF?
4. What happened to the frequency as the inductance changed and the capacitance was held at 14 μF?

SERIES LC CIRCUITS

One thing to remember in any series circuit is its current. The current is the same in all series circuits, which means in this series LC circuit that the current through the inductor also appears to flow through the capacitor. In most instances, the series LC circuit is analyzed as part of the series RCL circuit. This is because at resonance the series LC circuit appears as a short circuit to the frequency it is tuned to. When $X_L = X_C$ in a series LC circuit, the impedance is the resistance of the coil and nothing more. In a purely inductive coil (with no resistance in the coil) with a purely capacitive capacitor, the theoretical conditions would exist. However, since we have not been able to arrange for a purely inductive or a purely capacitive type of device, we have to assume there will be some resistance in a series LC circuit, even at resonance. It does, then, have a *minimum* amount of opposition to the frequency to which it is resonant. This minimum amount of opposition will cause an *infinite*, or—in practical terms—maximum, amount of current.

Table 9-1. Conditions Existing in a Series LC Circuit

	Below Resonance	At Resonance	Above Resonance
X_L	Less than X_C	Equal X_C	Greater than X_C
Z of series circuit	Greater than minimum. Appears to be capacitive.	At minimum value. Appears to have only resistance.	Greater than minimum. Appears to be inductive.
Current in series circuit (same in all branches)	Less than at resonance.	Maximum at resonance.	Less than at resonance.
Voltages in series circuit	Sum of voltage drops tends to equal the applied voltage.	Maximum voltage drops across both components.	Sum of voltage drops tends to equal the applied voltage.
X_C	Greater than X_L	Equals X_L	Less than X_L

The best use for this type of circuit is as a filter or trap. It will trap out or bypass the unwanted frequency. However, if you resonate a capacitor and coil on a power line frequency such as 60 hertz, then the two components present a short circuit and can draw very high currents. This is one reason why this type of circuit needs a resistance inserted for protection of the power source and for current limitation purposes.

The best circuit analysis for the series LC circuit is found in the series RCL circuit. Take a look at it and eliminate the effect of the resistance in the circuit, and you will have the results of the series LC in most instances. Table 9-1 sums up the conditions and actions of the series LC circuit.

PARALLEL LC CIRCUITS

In our previous discussion of inductance and capacitance it was shown that a parallel RL circuit or a parallel RC circuit differed from its respective series circuits in that the voltage in the circuit is the same for all components, the currents are 90° out of phase, and the resultant line current either lags or leads the applied voltage by some angle less than 90°. In addition, it was shown that the line current is greater than the current in either branch, and that the total impedance is less than the impedance of either branch.

Unlike the series reactive circuits, an increase in the resistance of a parallel circuit lessens the current through that branch and increases the relative effectiveness of the inductance or capacitance, resulting in a more reactive circuit as the phase angle θ approaches 90°. The limiting condition in this instance would be a resistance of infinite ohms, effectively opening that branch and making current and voltage across the reactive element 90° out of phase. On the other hand, a decrease in the value of the resistance in a parallel circuit causes an increase in current in that branch and decreases the relative effectiveness of the inductance or capacitance, resulting in a more resistive circuit as θ approaches 0°. The limiting condition in this instance would be a resistance of zero ohms, directly short-circuiting the reactive element so that current and voltage are in phase in this short circuit.

When you bring together the parallel RL and RC circuits into a parallel RCL circuit, new relationships, complicated by the special effect of L and C at resonance, are introduced among the elements. Therefore, by ignoring for the moment the effect of resistance in the circuit, the characteristics of the LC circuit alone may be considered. A of Fig. 9-1 shows the schematic diagram of an LC circuit, in which the inductive reactance is 100 ohms, the capacitive reactance 50 ohms, and the applied voltage 300. The graphical representation of the voltages and currents in this circuit is shown in B of Fig. 9-1. It will be noted that since the applied voltage is the same across each component, the voltage may be used as a point of reference. Then the current through the inductance is plotted as lagging this voltage by 90°, and the current through the capacitance as leading this voltage by 90°. The two currents are then seen to be 180° out of phase.

The current through the inductance is I_L:

$$I_L = \frac{E}{X_L} = \frac{300\text{V}}{100\Omega}$$

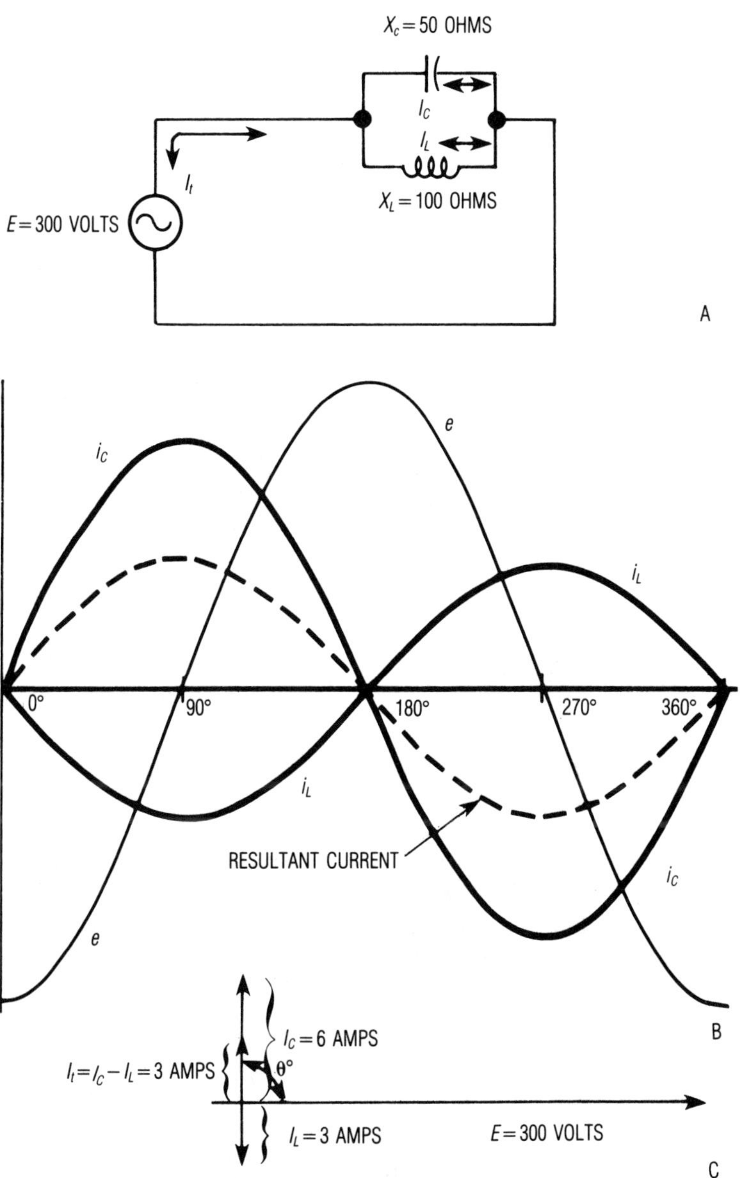

$X_c = 50$ OHMS
I_C
I_L
I_t
$X_L = 100$ OHMS
$E = 300$ VOLTS
A
i_C
e
i_L
0°
90°
180°
270°
360°
i_L
RESULTANT CURRENT
i_C
e
B
$I_C = 6$ AMPS
$I_t = I_C - I_L = 3$ AMPS
θ°
$I_L = 3$ AMPS
$E = 300$ VOLTS
C

Fig. 9-1

$$I_L = 3 \text{ amperes}$$

The current through the capacitance is I_C:

$$I_C = \frac{E}{X_C} = \frac{300\text{V}}{50\Omega}$$

$$I_C = 6 \text{ amperes}$$

Since these two currents are 180° out of phase, they may be added algebraically to find the total line current:

$$I_T = I_C - I_L$$

$$I_T = 3 \text{ amperes}$$

The resultant total line current is 3 amperes (see the dotted sine wave curve of B of Fig. 9-1). This current leads the applied voltage by 90°, and the circuit appears to the source as a capacitive circuit. The effect of the inductance is completely canceled out by the greater effect of the capacitance. C of Fig. 9-1 is the vector diagram that shows this result.

The total impedance of this type of parallel circuit is of particular interest. In all previous discussions, it was found that the total impedance of any type of parallel a.c. or d.c. circuit was always less than the impedance of any branch. But in the circuit in Fig. 9-1:

$$Z_T = \frac{E}{I_T} = \frac{300\text{V}}{3\text{A}}$$

$$Z_T = 100 \text{ ohms}$$

A total impedance of 100 ohms for this circuit is a value greater than X_C (50 ohms) and, in this case, equal to X_L (100 ohms). This result might have been obtained from the familiar formula for the impedance of a parallel circuit:

$$Z_T = \frac{X_L \times X_C}{X_L + X_C}$$

Substituting values:

$$Z_T = \frac{100 \times (-50)}{100 + (-50)}$$

Since X_L and X_C are also 180° out of phase, they may be added algebraically. The minus sign indicates capacitive reactance. This is usually taken as the sign opposite to that of inductive reactance. Then: $Z_T = -100$ ohms.

From the formula,

$$Z_T = \frac{X_C \times X_C}{X_L + X_C}$$

it is interesting and important to note that at resonance (when X_L is equal to X_C), the denominator of the equation is equal to zero and the total impedance approaches an infinite amount. Of course, L and C can never offer pure reactance to a circuit, and so it may be said that at resonance a parallel LC circuit offers maximum impedance to the applied voltage and line current falls to a minimum. On either side of resonance, the total impedance of the circuit falls off rapidly from its maximum (theoretically infinite value), and current rises as the impedance falls. A further examination of the formula will reveal that at the point at which one reactance is twice the value of the other, the total impedance of the circuit is equal to the larger reactance. As the difference between the values increases, the total impedance falls to a value between X_L and X_C, always greater than the smaller and less than the larger.

SERIES RCL CIRCUITS

Previously we found that:

1. The voltage drop across a resistor is *in phase* with the current through it.
2. The voltage drop across an inductor *leads* the current through it by 90°.
3. The voltage drop across a capacitor *lags* the current through it by 90°.

Three basic relationships between voltage and current are shown as graphs and as vectors in Fig. 9-2. These vectors indicate phase angle by their direction, and their magnitude depends on the values chosen for a given circuit. This means that if I is the effective value of the current in either L, C, or R, then the magnitudes of the effective voltages drops are:

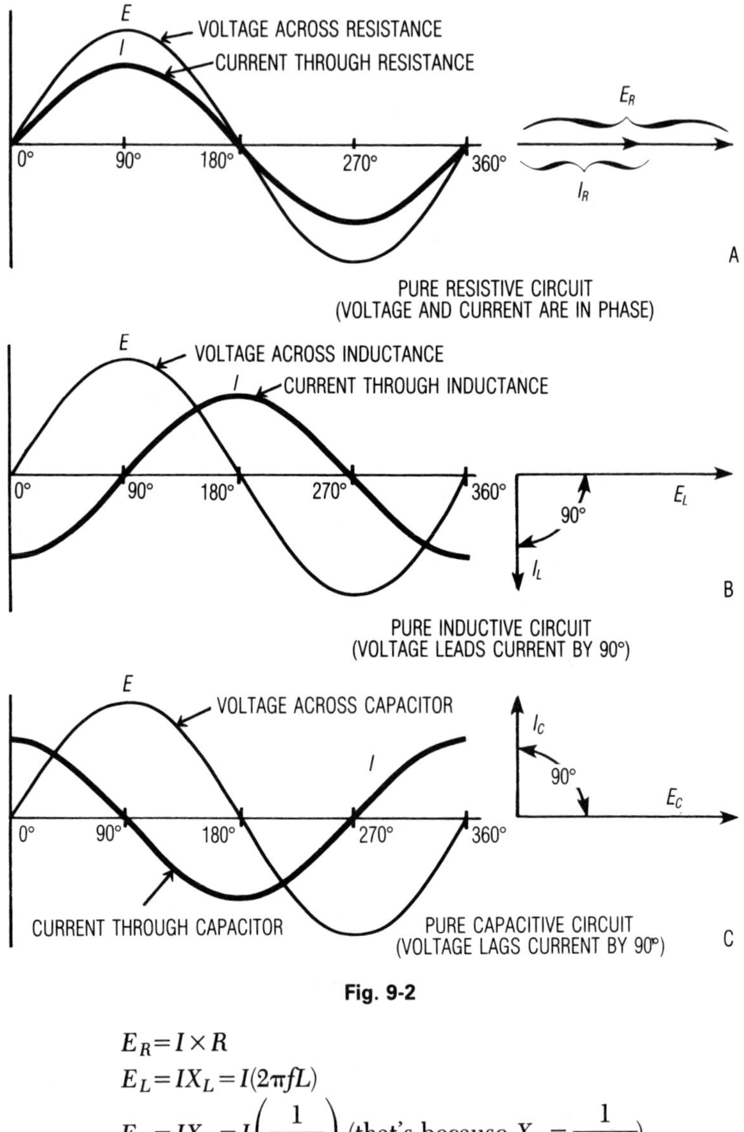

Fig. 9-2

$$E_R = I \times R$$
$$E_L = IX_L = I(2\pi fL)$$
$$E_C = IX_C = I\left(\frac{1}{2\pi fC}\right) \text{ (that's because } X_C = \frac{1}{2\pi FC}\text{)}$$

Current in an a.c. circuit varies with time. The voltage drops across the various elements also vary with time. But the same variation is not always present in each at the same time. Kirchhoff's second law (when applied to a.c. circuits) states that *at any instant* the sum of the voltage drops around a closed circuit equals the sum of the voltage rises, or the total applied voltage. In Fig. 9-3, a series RCL circuit is shown. The symbols e, e_R, e_L, and e_C denote *instantaneous* voltages. That produces:

$$e = e_R + e_L + e_C$$

This relationship is true for all values of time. But an instantaneous a.c. voltage cannot be determined by Ohm's law. Ohm's law holds for only the maximum, effective, or average values. An a.c. quantity is fully determined when its effective values and phase (in respect to some standard) are known. Therefore, the method of analysis by vectors, which show phase as direction and magnitude as effective value, can be used to add either sine voltages or sine currents.

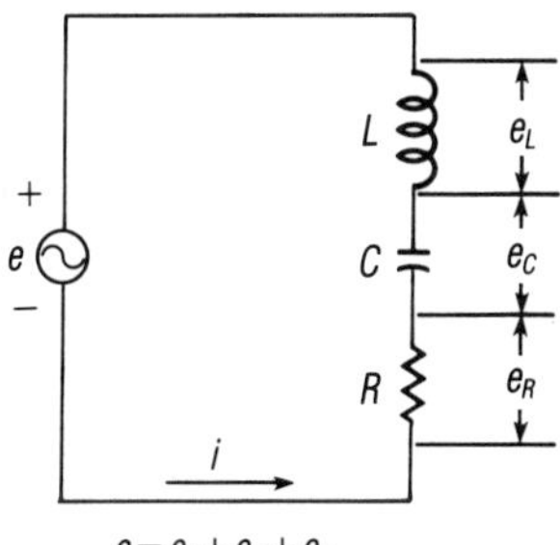

Fig. 9-3 $e = e_L + e_C + e_R$

Example 2

A 60-Hz, 100-volt generator is connected to a circuit having a resistance of 6 ohms in series with an inductive reactance of 8 ohms. See Fig. 9-4. What is the voltage across the components and the phase angle of the circuit?

1. What do you know and what is unknown?
2. *Known* | *Unknown*

Known	*Unknown*
$X_L = 8$ ohms	$E_L = ?$
	$E_R = ?$
$R = 6$ ohms	$I_T = ?$

3. Find the total current first so you can then multiply the current times the resistance and inductive reactance. This will produce the voltage drops across the inductor and the resistor.
4. In order to find the total current you can use:

$$E^2 = E_R{}^2 + E_L{}^2$$

5. Substitute the values known into the formula to obtain:

$$(100)^2 = (IR)^2 + (IX_L)^2$$
$$(100)^2 = 36I^2 + 64I^2$$
$$(100)^2 = 100I^2$$
$$I^2 = 100$$
$$I = 10$$

6. Now you know the total current. It is the same current in all parts of the circuit, so each component will have that current value through it.
7. That means:

$$E_R = I \times R$$
$$10 \times 6 = 60 \text{ volts}$$

8. That means:

$$E_L = I \times X_L$$
$$10 \times 8 = 80 \text{ volts}$$

9. Next, find the phase angle by using the tangent.
10. Tangent of the phase angle is equal to $\frac{E_L}{E_R}$.
11. Substitute and find the phase angle:

$$80/60 = 1.33333333.$$

12. Use the calculator or the trig tables to find the angle for 1.3333333 = 53.1°.

This tells you that the current in the circuit lags the applied voltage by 53.1°. C of Fig. 9-4 shows this graphically.

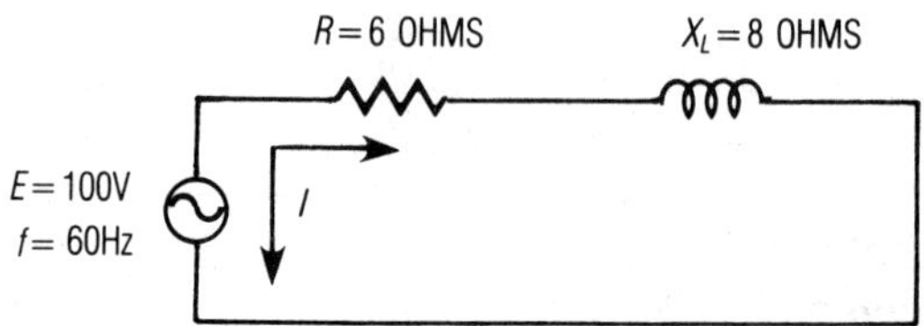
$R = 6$ OHMS
$X_L = 8$ OHMS
$E = 100V$
$f = 60Hz$
I

A

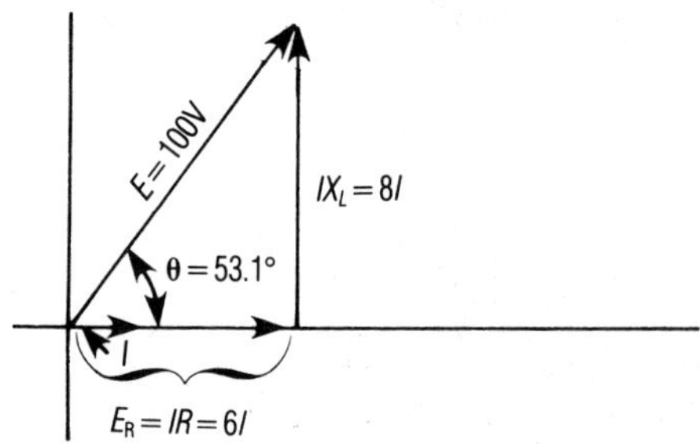
$E = 100V$
$IX_L = 8I$
$\theta = 53.1°$
I
$E_R = IR = 6I$

B

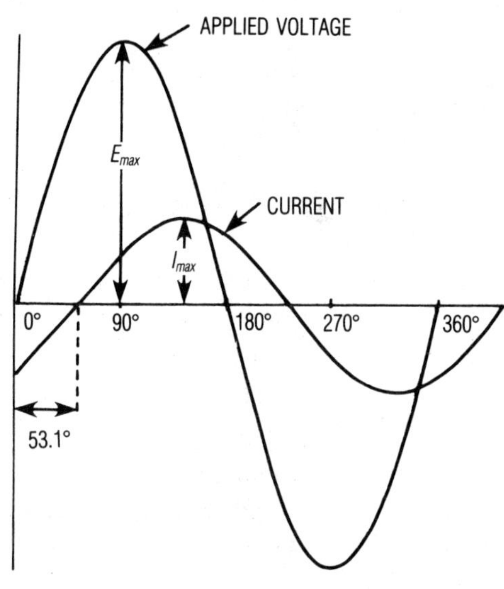
APPLIED VOLTAGE
E_{max}
CURRENT
I_{max}
0°
90°
180°
270°
360°
53.1°

C

Fig. 9-4

Example 3

Keep in mind that the current, voltages, and phase angle of a series RC circuit may be determined. A of Figure 9-5 shows the schematic diagram of a circuit of 3 ohms resistance and 4 ohms capacitive reactance connected to a 60-Hz, 300-volt a.c. source.

The vector sum of the voltage drops across the resistance and the capacitance equals the applied voltage. Since E_R is in phase with I, and E_C lags I by 90°, their vector sum, 300 volts, is the hypotenuse of a right triangle (B of Fig. 9-5). Then:

$$E^2 = E_R^{\,2} + E_C^{\,2}$$

Substituting equivalents produces:

$$(300)^2 = [(3I)^2 + (4I)^2] = 25I^2$$
$$300^2 = 25I^2$$

The square root of each:

$$300 = 5I$$
$$I = \frac{300}{5} = 60 \text{ amperes}$$

E_R, the voltage drop across the resistance, is IR, which is 60 times 3, which is equal to 180 volts. E_C, the voltage drop across the capacitance, is IX_C, which is 60 times 4 = 240 volts. The tangent of the phase angle θ is $\frac{E_C}{E_R} = \frac{240}{180} = 1.33$. From the trig table or using the calculator, you get 53.1°.

The current in this circuit leads the applied voltage by 53.1°. C of Fig. 9-5 presents these results graphically.

The Series RCL Circuit

When the three basic circuit elements of inductance, capacitance, and resistance are brought together in a single circuit, the voltage drops, current, and phase angle are determined by combining the methods just reviewed. A of Fig. 9-6 shows a series RCL circuit containing 6 ohms resistance, 8 ohms inductive reactance, and 16 ohms capacitive reactance connected to a 60-Hz, 300-volt source.

B and C of Fig. 9-6 are the graph and vector diagram for this

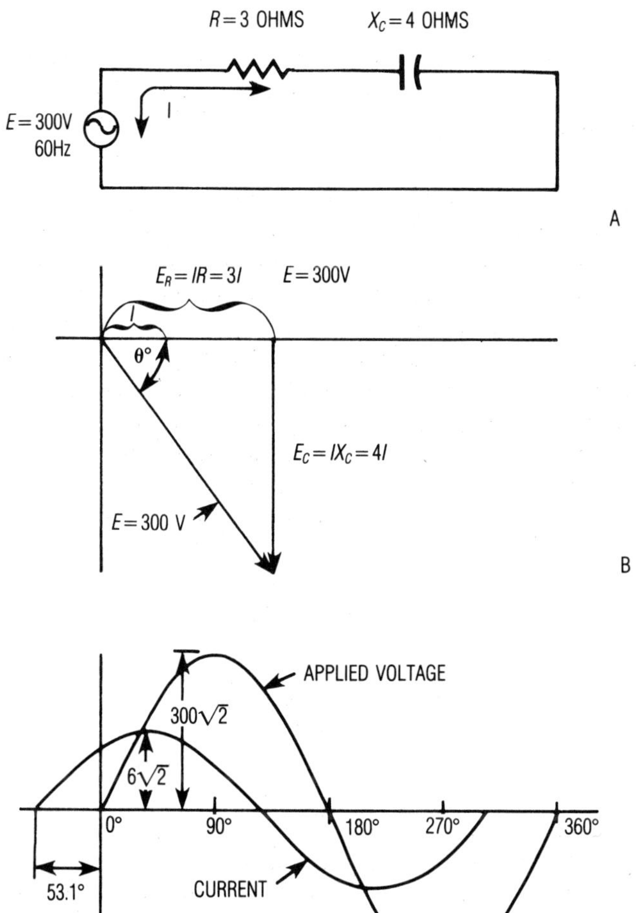

Fig. 9-5

circuit. Since current in a series circuit is the same through all components, it is the reference vector.

Then E_R is in phase with I, E_L leads I by 90°, and E_C lags I by 90°. E_C and E_L are 180° out of phase, and so their vector sum is merely the difference between the two. Since E_C is larger than E_L, resultant reactive voltage is $E_C - E_L$.

That means E_C is $I \times X_C$ and E_L is $I \times X_L$. Substitute and you get: $16 \times I$ and $8 \times I$, which produces $16I - 8I = 8I$.

This resultant voltage (represented by $8I$) lags the I by 90° because it has the direction of E_C, the larger vector. D of Fig. 9-6 is the resultant vector diagram. From the figure:

$$E^2 = (IR)^2 + (IX)^2$$
$$(300)^2 = [36I^2 + 64I^2] = 100I^2$$
$$300^2 = 100I^2$$

The square root of each is then taken to produce:

$$300 = 10I$$
$$I = \frac{300}{10} = 30 \text{ amperes}$$

The voltage drop across the resistance:

1. $E_R = I_R$, which is 30 times 6 or 180 volts
2. $E_L = IX_L$, which is 30 times 8 or 240 volts
3. $E_C = IX_C$, which is 30 times 16 or 480 volts.

Because the larger capacitive reactance cancels the inductive reactance, the circuit is capacitive, and so the current leads the applied voltage by the phase angle (θ). The tangent of θ is equal to $\frac{X_C - X_L}{R}$. That means when the proper values of 8 (found by subtracting X_C from X) and 6 (found by using the R value) are divided, they produce 1.33. The angle θ is 1.33, so you check out the tangent in the trig tables or you use the calculator to find the tangent of 1.33. It is equal to 53.1°.

The series RCL circuit shows the following important points:

1. The current in a series RCL circuit either leads or lags the applied voltage, depending on whether X_C is greater or less than X_L.
2. A capacitive voltage drop in a series circuit always *subtracts* directly from an inductive voltage drop.
3. The voltage across a single reactive element in a series circuit can have a greater effective value than that of the applied voltage.

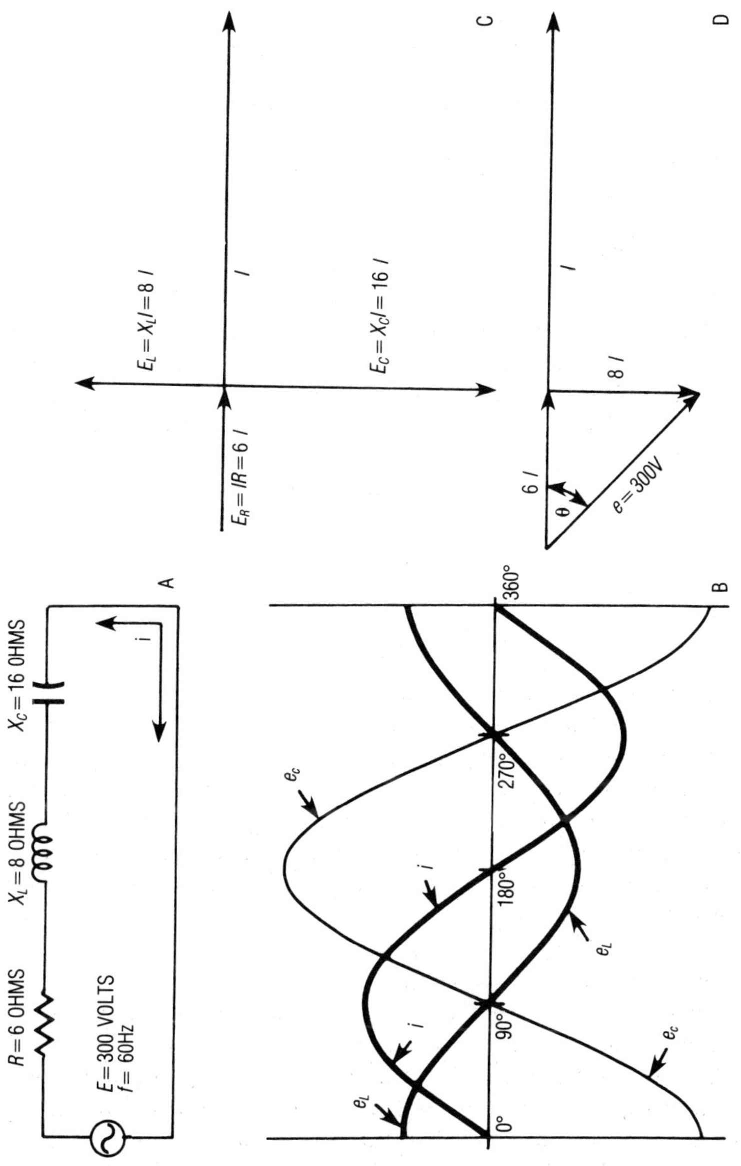

R=6 OHMS
X_L=8 OHMS
X_C=16 OHMS
E=300 VOLTS
f=60Hz
i
A
e_L
i
e_c
0°
90°
180°
270°
360°
B
E_L=X_LI=8 I
I
E_C=X_CI=16 I
E_R=IR=6 I
C
6 I
θ
8 I
e=300V
I
D

Fig. 9-6

As noted in 3 above and illustrated in the circuit of Fig. 9-6, E_C has an effective value of 480 volts. This is compared to the applied voltage of only 300. In d.c. circuits, this condition arithmetically and according to the voltage drops across any one of the component parts is always less than the applied voltage.

Analysis of RCL Circuits

The various relationships noted previously may now be generalized. The current, impedance, voltage drops, and phase angle of any series RCL circuit can be calculated. In A of Fig. 9-7, a schematic diagram of a series circuit containing an inductance L, a capacitance C, and a resistance R are shown. This circuit is connected to an a.c. voltage source of magnitude E and frequency f. B of Fig. 9-7 shows the initial vector diagram. The voltage drop across the resistance (IR) is drawn in phase with the current I. The voltage drop across the inductance is drawn above the zero axis. The voltage drop across the capacitance is drawn below the axis. C of Fig. 9-7 shows the resultant vectors. The voltage triangle (with base E_R and vertical side $E_L - E_C$) is shown in D of Fig. 9-7. The resultant voltage E, equal to the applied voltage, is the hypotenuse of the right triangle. That means:

$$E^2 = I^2R^2 + I^2(X_L - X_C)^2$$
$$E^2 = I^2(R^2 + (X_L - X_C)^2)$$

The square root:

$$E = \sqrt{R^2 + (X_L - X_C)^2}$$

And:

$$I = \frac{E}{\sqrt{R^2 + (X_L - X_C)^2}}$$

Also, since in any a.c. circuit I is equal to $\frac{E}{Z}$, then Z, the total impedance of any RCL circuit, is equal to the square root of the square of the resistance plus the square of the difference of the reactances present. Or:

$$Z = \sqrt{R^2 + (X_L - X_C)^2}$$

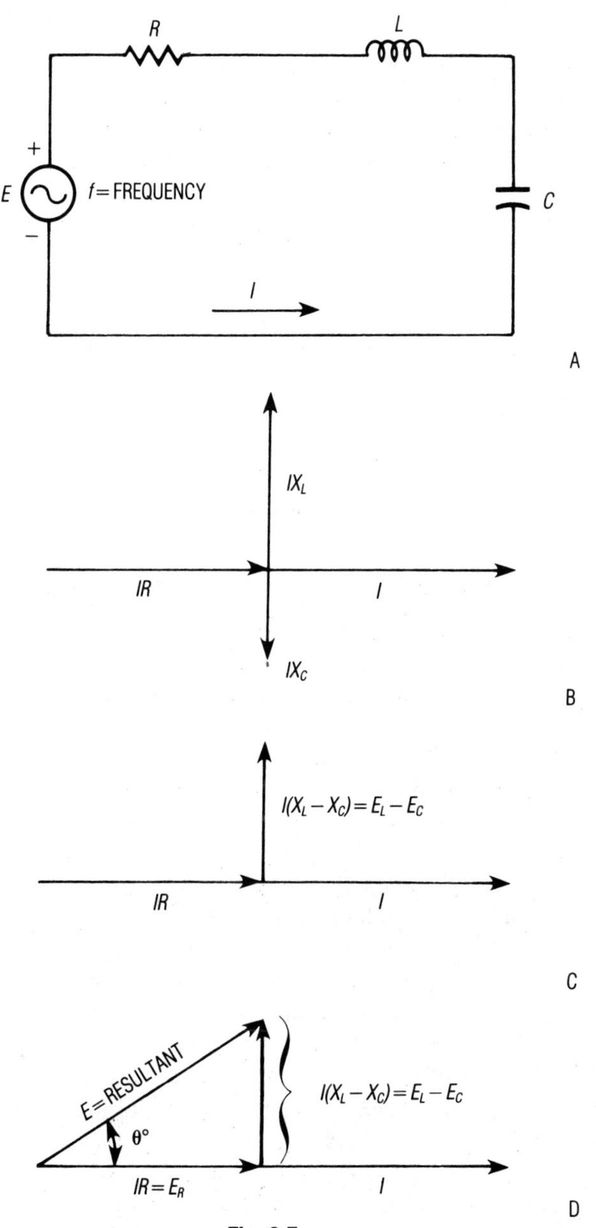
R
L
+
E
f = FREQUENCY
C
−
I
A
IX_L
IR
I
IX_C
B
$I(X_L - X_C) = E_L - E_C$
IR
I
C
E = RESULTANT
θ°
$I(X_L - X_C) = E_L - E_C$
$IR = E_R$
I
D

Fig. 9-7

From the voltage triangle, it is seen that the current lags the applied voltage by the angle θ. Then:

$$\text{Tan } \theta = \frac{E_L - E_C}{E_R}$$

By substitution and cancelation:

$$\text{Tan } \angle\theta = \frac{X_L - X_C}{R}$$

In a like manner, the cosine of θ is equal to $\frac{E_R}{E}$, or:

$$\text{Cos } \angle\theta = \frac{R}{Z}$$

Then:

$$Z = \frac{R}{\text{Cos } \theta}$$

These results may be written in another form. Since X_L is equal to $2\pi fL$, and X_C to $1/2\pi fC$, then:

$$Z = \sqrt{R^2 + \left(2\pi fL - \frac{1}{2\pi fC}\right)^2}$$

In all of the preceding calculations, the values of I and E are the effective or *rms* values of current and voltage.

From the general formulas given above it may be seen that if the inductive reactance X_L is greater than the capacitive reactance X_C, the inductive voltage drop is greater than the capacitive voltage drop and their sum is inductive. That is, the voltage leads the current by 90°. If X_C is greater than X_L, the reverse is true—that is, the sum is capacitive and the voltage lags the current by 90°. Finally, if the inductive reactance equals the capacitive reactance, the sum is zero and a very interesting phenomenon called *resonance* occurs.

In the case of resonance, the inductive and capacitive voltage drops are equal in magnitude and 180° out of phase. That means their vector sum is 0. Physically, this means that at every instant of time, the voltage drop across the capacitor e_C added to the voltage drop across the inductor e_L gives a resultant voltage of zero. Expressed as an equation, it becomes:

$$e_C + e_L = 0$$
$$e_C = -e_L$$

These results are shown graphically in A of Figure 9-8. Thus, a series RCL circuit is said to be resonant when X_L is equal to X_C. Then, since no reactive voltage is present, the impedance of the circuit is equal to the d.c. resistance of the circuit and current is limited by this resistance alone.

The phase angle θ has a tangent of 0.
The phase angle θ has a cosine of 1.
The phase angle θ has an angle of 0.

This means the current in the circuit is in phase with the applied voltage. The vector for this condition is shown in B of Fig. 9-8. It

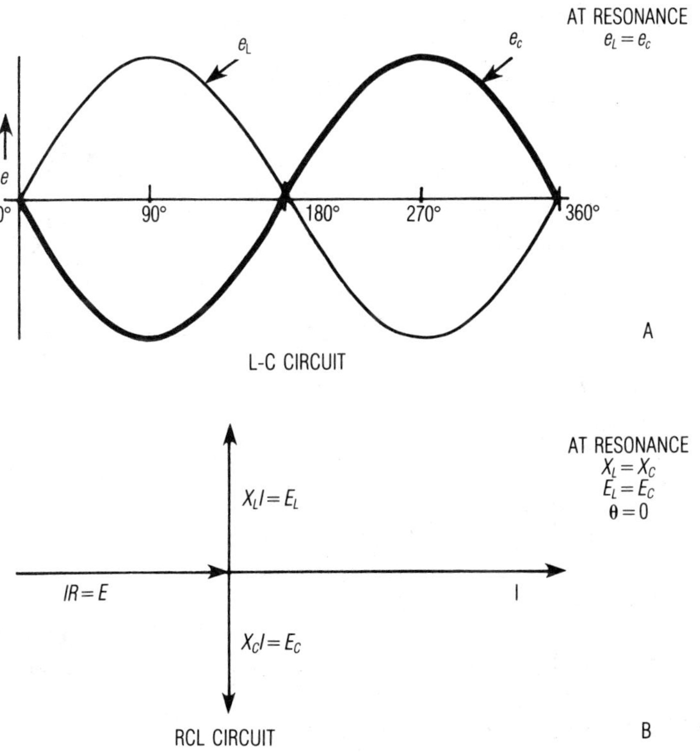

Fig. 9-8

should be noted, however, that although the vector sum of the reactive voltages is 0, the voltage across either element will be very high, since the voltage drop across the element is determined by its reactance and the current through it. Current in a resonant circuit is a maximum because it is limited only by *R*, and therefore the voltage drops across the reactive elements are also maximum and may be far in excess of the applied voltage.

A series RCL circuit may then appear to its a.c. source as any one of the following types:

1. An RL circuit ($X_L > X_C$): Inductive reactance is greater than the capacitive reactance.
2. An RC circuit ($X_C > X_L$): Capacitive reactance is greater than the inductive reactance.
3. An R circuit ($X_L = X_C$): Inductive reactance is equal to the capacitive reactance.

The resonant frequency of the RCL circuit may be determined by reference to individual reactance, since both depend on frequency. Inductive reactance is equal to $2\pi fL$, and capacitive reactance is equal to $\frac{1}{2\pi fC}$. Then, since at resonance X_L is equal to X_C:

$$2\pi fL = \frac{1}{2\pi fC}$$

Solving for *f*:

$$f^2 = \frac{1}{4\pi^2 LC}$$

$$f = \frac{1}{2\pi\sqrt{LC}}$$

This is the frequency at which resonance occurs in a series RCL circuit. Regardless of the values of *L* and *C*, the circuit is resonant at some frequency. If the source frequency is fixed, a circuit may be designed to resonate to that frequency.

PROBLEMS

1. Find the values and fill in the boxes below for a *series RCL circuit* that has an inductor with 10 henrys, a capacitor of 10 microfarads and a resistance of 10 ohms.

	Frequency (Hz)	X_L (Ω)	X_C (Ω)	I_L (A)	I_C (A)	$\angle\theta$ (degrees)	Applied Voltage (V)
a.	60						120
b.	50						220
c.	400						120
d.	400						110
e.	25						115
f.	25						230

2. Find the values and fill in the boxes for a *series RCL circuit* that has an inductor of 5 henrys, a capacitor of 100 microfarads, and a resistance of 1,000 ohms.

	Frequency (Hz)	X_L (Ω)	X_C (Ω)	I_L (A)	E_L (V)	E_C (V)	E_R (V)	Applied Voltage (V)
a.	60							120
b.	60							240
c.	60							440
d.	60							880
e.	50							220
f.	50							240
g.	25							110
h.	25							220

PARALLEL RCL CIRCUITS

The parallel RCL circuit may be seen as nothing more than a parallel LC circuit with an added resistance *R* in parallel with the inductor and capacitor in the circuit. A of Fig. 9-9 shows this type of circuit. At any nonresonant point where inductive reactance is not equal to capacitive reactance, the currents through the inductor (L) and the capacitor (C) will be unequal and 180° out of phase. The resultant current will therefore be either predominantly inductive or capacitive, depending on the branch offering the least opposition to the applied voltage. The generator looking into this circuit, then, would see either a capacitance or an inductance in parallel with the resistance R. The resultant line current is the vector sum of the reactive current and the resistive current, as though the circuit comprised a single reactance and a resistance.

Example 4

For instance, circuit A of Fig. 9-9 shows an input voltage of 300, an inductive reactance of 75 ohms, a capacitive reactance of 50 ohms, and a resistance of 100 ohms. B of Fig. 9-9 is the graphical representation of the currents in this circuit. It can be seen that the reactive currents are 180° out of phase with each other and 90° out of phase with the current through the resistance, which is in phase with the applied voltage. Then:

$$I_L = \frac{300\text{V}}{75\ \Omega} = 4\text{ A}$$

$$I_C = \frac{300\text{V}}{50\ \Omega} = 6\text{ A}$$

$$I_R = \frac{300\text{V}}{100\ \Omega} = 3\text{ A}$$

The total reactive current I_X is equal to 2 amperes ($I_C - I_L$). C of Fig. 9-9 shows the vector diagram of the currents. Thus the resultant line current is the vector sum of the reactive current (capacitive) and the resistive current:

1. $I_T = \sqrt{I_X{}^2 + I_R{}^2}$
2. $I_T = \sqrt{2^2 + 3^2}$
3. $I_T = \sqrt{4 + 9}$
4. $I_T = \sqrt{13} = 3.605551275\text{A}$

Then Z_T, the total impedance in the circuit, is equal to $\frac{E}{I_T}$:

$$Z_T = \frac{300\text{V}}{3.605551275\text{A}} = 83.20502944\Omega$$

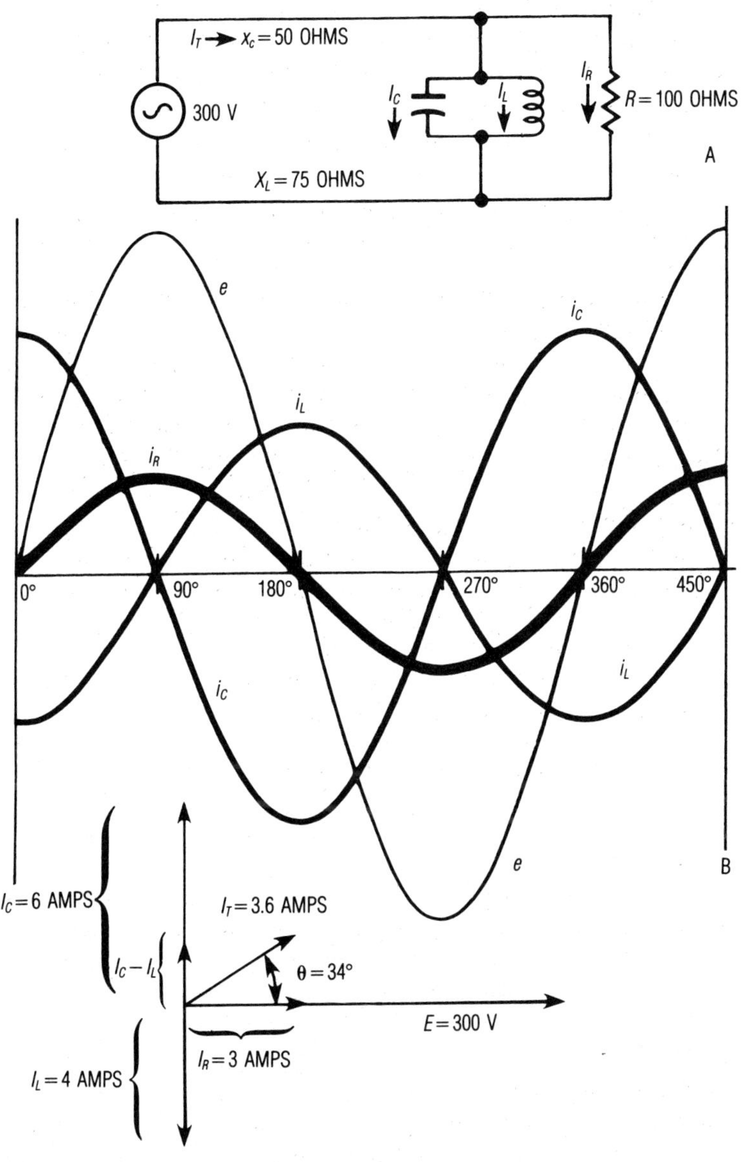

I_T → $X_C = 50$ OHMS
300 V
I_C
I_L
I_R
$R = 100$ OHMS
A
$X_L = 75$ OHMS
e
i_C
i_L
i_R
0°
90°
180°
270°
360°
450°
i_C
i_L
e
B
$I_C = 6$ AMPS
$I_T = 3.6$ AMPS
$I_C - I_L$
$\theta = 34°$
$E = 300$ V
$I_R = 3$ AMPS
$I_L = 4$ AMPS
C

Fig. 9-9

The cosine of the phase angle θ that the line current leads the applied voltage is $\frac{Z}{R}$:

$$\text{Cos } \theta = \frac{83.20502944}{100} = 0.8320502944$$
$$\angle\theta = 33.69006752°$$

Resonance

At the point of resonance, the two reactive currents effectively cancel each other, leaving only the current through the resistance as the total current of the circuit. That means:

$$I_C - I_L = 0$$
$$I_T = I_R$$
$$I_T = \frac{E}{R}$$
$$Z_T = R$$
$$\text{Cos } \angle\theta = \frac{Z}{R} = 1$$
$$\angle\theta = 0°$$

Line current and voltage are effectively in phase across the resistance. Again, it should be noted that a resistance in parallel to an LC circuit tends to destroy the effect of the reactive components, as is evident when the RL and RC circuits are studied. At resonance, the impedance Z of the circuit is that of the resistance R. This destroys the chief characteristic of a parallel resonant circuit—its maximum impedance to the line. It does have uses, though. The resistor inserted across a tuned resonant LC circuit has a tendency to broaden the bandpass of that particular resonant circuit. This characteristic is used widely in FM and TV receivers in the intermediate frequency (IF) amplifier stages.

Example 5

What is the phase angle, total current, and impedance of a parallel RCL 120-volt a.c. circuit with a resistor of 1,000 ohms, an inductor with 1,884 ohms inductive reactance, and a capacitor with 1,000 ohms of capacitive reactance?

1. What do you know?
   a. $X_L = 1{,}884$ ohms
   b. $X_C = 1{,}000$ ohms
   c. $R = 1{,}000$ ohms
   d. $E_A = 120$ V a.c.
2. What do you need to find?
   a. $Z = ?$
   b. $I_T = ?$
3. You need to find I_L, I_C, and I_R so you can obtain I_T. Therefore:
   a. $I_R = \frac{E}{R} = \frac{120}{1{,}000} = 0.120$ A
   b. $I_C = \frac{120}{1{,}000} = 0.120$ A
   c. $I_L = \frac{120}{1{,}884} = 0.0636942675$ A
4. You need I_T to find Z, so use the formula for I_T:

$$I_T = \sqrt{I_R{}^2 + (I_C - I_L)^2}$$

5. Substitute values into the formula and solve.

$$\sqrt{(0.120)^2 + (0.120 - 0.0636942675)^2}$$

$$\sqrt{(0.120)^2 + (0.0563057324)^2}$$

$$\sqrt{0.0144 + 3.170335509 \times 10^{-3}}$$

$$\sqrt{0.0175703355}$$

$$0.1325531422 \text{ A}$$

6. Next, you need to find Z:

$$Z = \frac{E_A}{I_T}$$

7. Substitute and solve:

$$Z = \frac{120}{0.1325531422} = 905.2972869\Omega$$

8. Now that you have Z and I_T, you can find the phase angle because:

$$\text{Cos } \angle\theta = \frac{I_R}{I_T}$$

$$\text{Cos } \angle\theta = \frac{0.12}{0.1325531422} = 0.9052972869$$

$$\angle = 25.13664754°$$

PROBLEMS

Find the phase angle, the total current and impedance of the parallel RCL circuit below with the given values:

E_A (V)	Z (Ω)	I_T (A)	$\angle\theta$ (degrees)	R (Ω)	X_L (Ω)	X_C (Ω)
100				100	50	25
50				100	25	50
25				2000	2500	2500
10				5	10	20

POWER IN A.C. CIRCUITS

In d.c. circuit analysis the amount of power absorbed by a resistor or by the resistance of a circuit is determined easily and simply by Joule's law:

$$P = I^2R$$

P = power absorbed in watts
I = total current in amperes
R = resistance of the circuit in ohms

Since the voltage drop across a resistor R is equal to IR, the formula above may be written:

$$P = IR \times I = EI$$

This expression for determining power in d.c. circuits is a general one and can be applied to any passive network (one which contains no internal batteries or other source of energy). Figure 9-

10 shows such a network. The power absorbed by the network is equal to the voltage applied to it multiplied by the current that flows in it, or $P = EI$.

In dealing with a.c. circuits the determination of power is a more complicated process. Since both current and voltage vary with time, the product of e times i (lowercase letters are used to indicate the voltage and current when instantaneous values are represented)

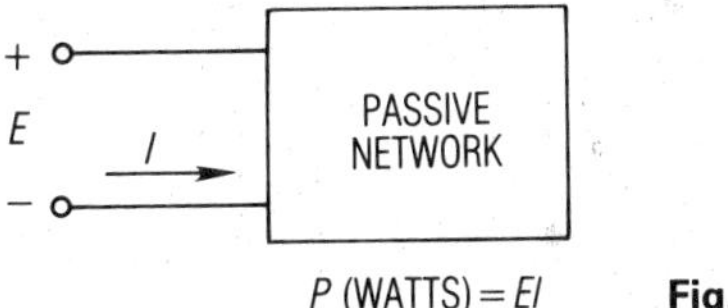

Fig. 9-10

is also a function of time and is called the instantaneous power p. In general, however, current and voltage in a.c. networks are out of phase by some angle θ, as previously discussed in RCL circuits. In Fig. 9-11, the conditions in such a circuit with a phase shift of 0° is illustrated by means of graphs for the current, voltage, and instantaneous power. These graphs reveal several important characteristics of instantaneous power:

1. The graph of power is of double frequency variation. That is, the instantaneous power p goes through two cycles during one period of voltage or current.
2. The power curve has positive loops and negative loops. Thus, for a part of the cycle, p is negative, which must be interpreted to mean that energy is being returned to the generator source during this time. This is a very important fact to remember since it indicates that, in an a.c. circuit, energy is delivered to the circuit by the source for parts of the cycle and returned to the source by the circuit for the remainder of the cycle. Accordingly, if in one cycle the amount of energy returned is equal to the amount delivered by the generator, the net power absorbed by the circuit is 0.

The area under the positive loops of the power curve measures the energy delivered to the load, and the area under the negative loops measures the energy returned to the source. In view of these facts, the following generalizations can be made:

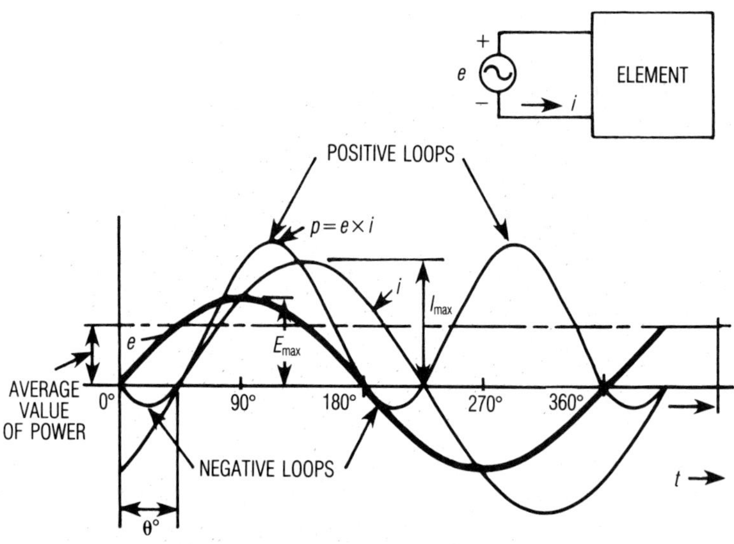

Fig. 9-11

1. If in one cycle (360°) the area under the positive loops of the *p* curve is greater than the area under the negative loops, the net energy delivered by the generator to the load is positive.
2. If in one cycle (360°) the area under the positive loops equals the area under the negative loops, the net energy delivered by the generator to the load is 0.
3. If in one cycle (360°) the area under the positive loops is less than the area under the negative loops, the generator is absorbing a net amount of energy.

Therefore, it may be stated that:

1. If the net energy delivered by the generator to the load is positive, the circuit contains some resistance, since energy is dissipated in a resistor in the form of heat, which cannot be returned to its source.

2. If the net energy delivered by the generator to the load is 0, the circuit must contain purely reactive elements, since no energy is used up in the network.
3. If the net energy delivered by the generator to the load is negative, the generator is absorbing energy and the network contains its own generator, since it is actually delivering power back to the source.

Figure 9-12 shows these characteristics with graphs of voltage, current, and power for the three general types of a.c. circuits: capacitive, inductive, and resistive. In A and B of Fig. 9-12, the area under the positive loops equals the area under the negative loops. The capacitor shown in the schematic of A charges during the first quarter-cycle, and discharges back into the generator during the next quarter-cycle. The same action is repeated during each half-cycle of the generator voltage. B illustrates the same action for a pure inductance. C shows that in the case of a pure resistance, the power loops are always positive—that is, the resistor R absorbs energy completely, returning none of it to the source.

The behavior of the reactive elements in the circuits above should not be surprising because, as was pointed out earlier, both capacitors and inductors are elements capable of storing energy without loss. For example, the energy used to charge a capacitor can be completely regained by allowing it to discharge through a resistor; the electrostatic energy of the capacitor is thus converted into heat energy. Also, the energy used to develop the magnetic field about an inductor can be regained by allowing the current induced in it by the collapsing field to discharge through a resistor. Figure 9-13 illustrates how the discharging of a reactive element is accomplished. In A, the capacitor C is initially charged. When the switch is closed, C discharges through the resistance R. In B, the initial current through the inductance L stored energy by building up a magnetic field; when the switch is closed, the energy stored in the field produces an induced voltage in the inductance and, accordingly, a current through the resistance R, thus dissipating the energy as heat. The limiting resistor R_0 is placed in the circuit to prevent the battery from being short-circuited when the switch is closed.

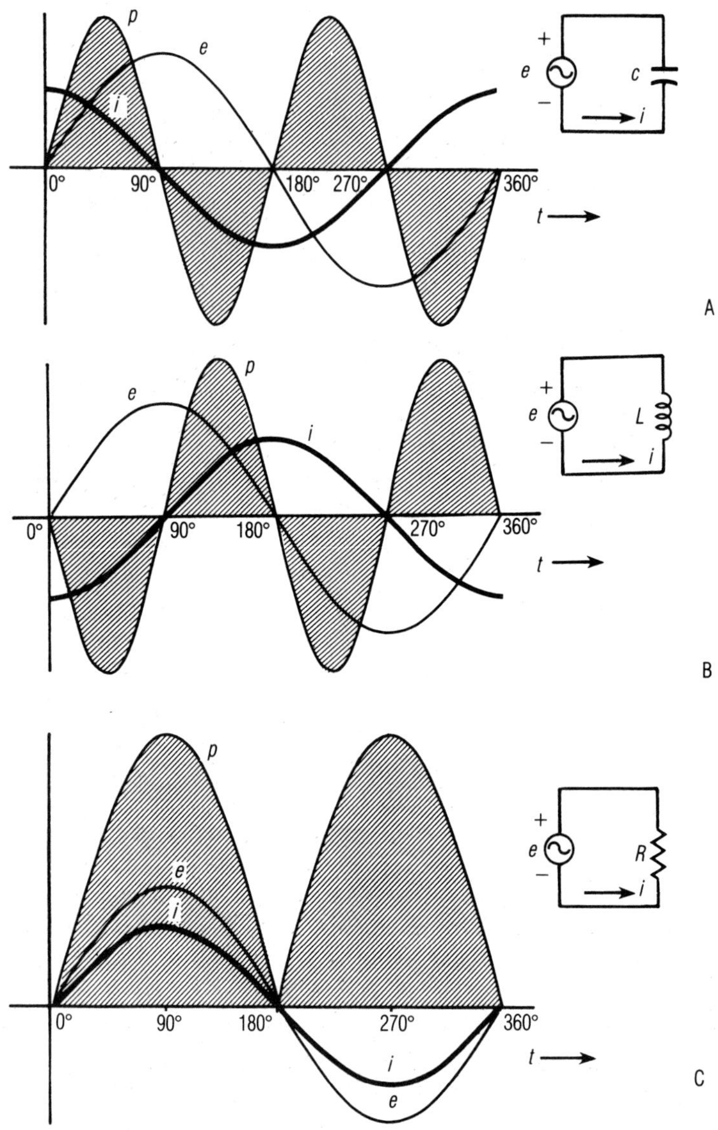

p
e
i
0°
90°
180°
270°
360°
t
e
C
+
−
i
A
p
e
i
0°
90°
180°
270°
360°
t
e
L
+
−
i
B
p
e
i
0°
90°
180°
270°
360°
t
i
e
e
R
+
−
i
C

Fig. 9-12

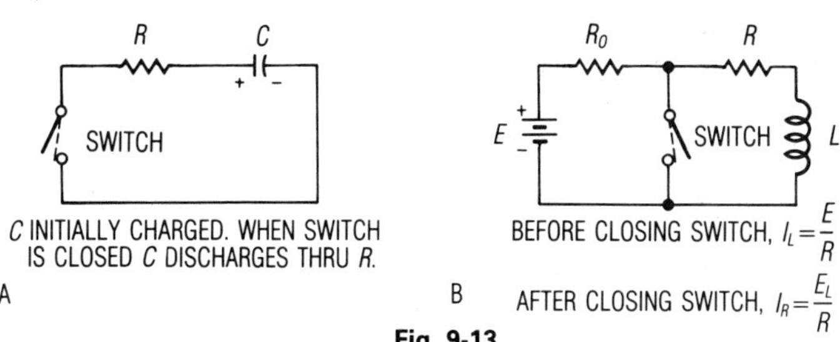

Fig. 9-13

True Power and Apparent Power

It is seen from these discussions of instantaneous power that in any a.c. circuit containing reactive elements, the only power actually dissipated is the power absorbed by the resistance of the circuit. A reactive circuit, however, appears to consume large quantities of power, as may be inferred from the area of the positive loops of Fig. 9-12. Thus it is important to note that even though the generator receives back certain amounts of energy from the load, it may supply large amounts to the load. This power, which the generator must deliver (regardless of the return), is called the apparent power; and, as in any ac circuit, is equal to the product of the effective value of the voltage and the current. Hence:

$$P \text{ (apparent power)} = EI \text{ (measured in \textit{volt-amperes}, VA)}$$

Apparent power is different from the actual power consumed by the load. That which is consumed is the *true* power and is the energy *absorbed by the resistance* of the circuit. True power is defined as the power absorbed by a circuit over a period of one cycle of the input voltage.

Note that the difference between the positive and negative power loops is the *difference* between the power delivered to the load and the power returned to the source. It is the power actually absorbed by the resistance of the circuit. Therefore, Joule's law expresses the actual true power consumed by any a.c. or d.c. circuit.

Thus, as we discussed previously:

$$P \text{ (true power)} = I^2R$$

But in a reactive a.c. circuit (one with either a capacitor or inductor or both), the current in the circuit is equal to $\frac{E}{Z}$. Then this makes, by substitution:

$$P=\frac{E}{Z}\times I\times R$$

Rewriting it in another form:

$$P=EI\times\frac{R}{Z}$$

However, the ratio $\frac{R}{Z}$ is known from the discussions of RCL circuits to be the cosine of the phase angle θ, or the angle between current and voltage in the circuit. That means true power is equal to EI multiplied by the cosine of θ. The cosine θ takes into consideration the phase angle and the amount of energy put back into the circuit by the reactive elements. The formula then becomes:

$$P=EI\cos\theta$$

E is the effective value of the voltage across the circuit
I is the effective value of the current in the circuit
θ is the phase angle between voltage and current
P is the average power absorbed by the circuit

The various derived formulas for apparent power follow:

$$P=EI$$
$$P=I^2Z$$
$$P=\frac{E^2}{Z}$$

And those for true power:

$$P=EI\cos\theta$$
$$P=I^2R$$

An examination of the formula for average power above reveals that if the phase angle θ equals 90°, its cosine is 0 and the actual power absorbed by the circuit is 0. Thus, the phase angle of 90° means that the circuit is purely reactive and returns as much power as it receives. If the phase angle is 0°, its cosine is 1.0, the circuit is purely resistive, and all the power produced by the source is absorbed by the load. The cosine θ then varies from 0 to 1 as the phase angle varies from 90° to 0°.

In the following, the effect of the phase angle on power is shown by the three circuits of Fig. 9-14, to each of which a 300-volt, 60-Hz a.c. current is applied.

The circuit of A of Fig. 9-14 comprises an inductive reactance (X_L) of 100 ohms and a capacitive reactance (X_C) of 200 ohms. Since these reactances are *opposed*, the effective reactance is 100 ohms *capacitive*, and the current in the circuit is 3 amperes. The apparent power in the circuit is:

$$P = EI$$

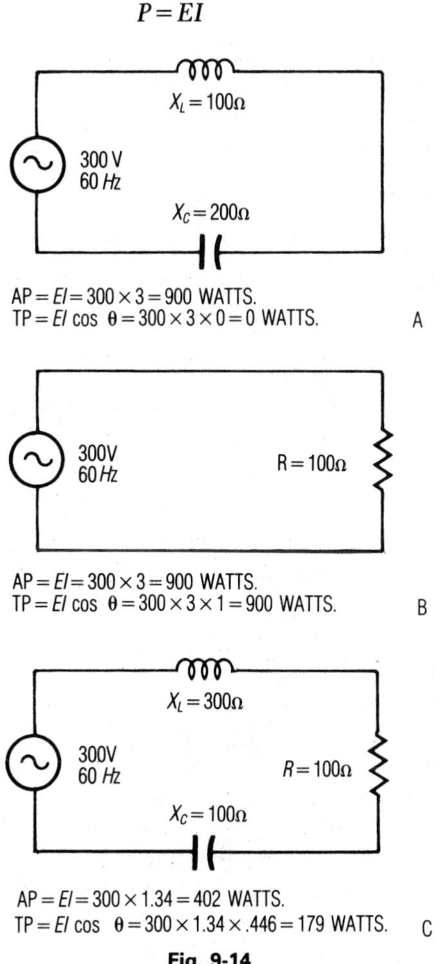

Fig. 9-14

By substituting, you have:

$$P = 300 \times 3$$
$$P = 900 \text{ watts}$$

The phase angle of this *purely capacitive* circuit is 90° and current is leading the voltage. Then the *true power* is found by $P = EI \cos \theta$. By substituting, you have $P = 300 \times 3 \times \cos 90°$. That produces $P = 900 \times 0$. True power then is equal to 0 watt. That means no power is consumed in this circuit, but the generator must supply 900 watts of power, all of which it receives back.

In B of Fig. 9-14 a pure resistance of 100 ohms is connected across the generator. The current in the circuit is again 3 amperes, and since the voltage and current are in phase, the phase angle is 0°. The apparent power (AP) is $P = EI$. By substituting, you have $P = 300 \times 3$, or 900 volt-amperes.

The true power (TP) is $P = EI \cos \theta$. By substituting, you have $P = 300 \times 3 \times \cos \theta$. The true power then is $P = 900 \times 1 = 900$ watts.

Thus the apparent power (AP) and the true power (TP) in an a.c. resistive circuit are the same.

In C of Fig. 9-14 an RCL circuit is shown with inductive reactance (X_L) of 300 ohms, capacitive reactance (X_C) of 100 ohms, and resistance (R) of 100 ohms. The impedance (Z) of this circuit is:

$$Z = \sqrt{R^2 + (X_L - X_C)^2}$$

Substituting:

$$Z = \sqrt{(100)^2 + (200)^2}$$
$$Z = \sqrt{50{,}000} = 223.6067977 \text{ ohms}$$
$$I = \frac{E}{Z}$$

Substituting:

$$I = \frac{300}{223.6067977}$$
$$I = 1.341640787 \text{ amperes}$$

The phase angle of this effectively inductive circuit may be determined from its cosine:

$$\text{Cos } \angle\theta = \frac{R}{Z}$$

Substituting:

$$\text{Cos } \theta = \frac{100}{223.6067977}$$

$$\text{Cos } \angle\theta = 0.4472135956$$

$$\angle\theta = 63.43494881°$$

The true power actually consumed by the circuit is:

$$P = EI \cos \theta$$

Substituting:

$$P = 300 \times 1.341640787 \times 0.4472135956$$

$$P = 180 \text{ watts}$$

The apparent power in the circuit is:

$$P = EI$$

Substituting:

$$P = 300 \times 1.341640787$$

$$P = 402.5 \text{ watts}$$

That means 180 watts of power are consumed in this circuit, but the generator must supply 402 watts, 222 watts of which are returned to the generator by the effective *reactive* element

The generator must produce this higher wattage, so you have to pay for it. If you are using a motor (inductive load), for instance, it draws more power (figured simply by the kilowatt meter on the house by multiplying voltage times current) and it is figured in volt-amperes, not watts. The phase angle is not taken into consideration by the kilowatthour meter on the wall. However, large industries that use many motors or large numbers of fluorescent lamps may want to use a means of bringing the true power into view rather than the apparent power. This is where the power factor comes into play. The following will give you more details on this important part of electricity and the electrical power field.

Power Factor

In reactive a.c. circuits, the relative amounts of apparent power and true power are an important consideration from the point of view of efficiency and circuit design. In the circuits previously shown in this chapter, it may be noted that the true power differs from the apparent power by the factor of the cosine of $\angle\theta$. Thus, the cosine θ determines the percentage of apparent power consumed as true power. The cosine of θ (which is the phase angle between current and voltage), then, is by extension a measure of the reactance present in the circuit. It is called the *power factor* of the circuit. That means:

$$\text{PF (Power Factor)} = \cos \angle\theta.$$

The power factor tells at a glance the relative power dissipated or consumed.

The concept of power factor may be understood from another point of view. Since the cosine of θ is equal to the ratio $\frac{R}{Z}$, both numerator and denominator of this expression may be multiplied by I^2. Then:

$$\text{Cos } \theta = \frac{I^2R}{I^2Z} = \frac{I^2R}{EI}$$

However, I^2R is equal to true power, and either I^2Z or EI is equal to the apparent power. Therefore, the cosine of θ is seen to be the ratio of true power to apparent power. This ratio is the power factor and is that decimal fraction between 0 and 1 which represents the amount of power actually used when compared to the power flowing in the circuit. Power factor, then, is:

$$\text{PF} = \frac{\text{True Power}}{\text{Apparent Power}}$$

Power factor is also equal to the cosine θ.

It is important to realize that a power factor close to 1 is generally to be desired for all reactive circuits using appreciable power. A low power factor means that there is a large discrepancy between the voltages and currents in the circuit and the small percentage of I and E needed to perform the work desired. The generator source and circuit elements, therefore, would have to be designed to produce and withstand the larger values. As the power factor approaches 1, the generator and circuit elements need be only slightly larger than their useful values. For example, in the circuit C of Fig. 9-14 the power factor is (values rounded):

$$PF = \frac{TP}{AP}$$

$$PF = \frac{180W}{402VA}$$

$$PF = 0.4477 = \cos\theta.$$

The current through all elements is 1.34 amperes. The voltage across the inductance E_L is equal to IX_L, and the voltage across the capacitance is equal to IX_C. Then:

$$E_L = 1.34 \times 300$$

$$E_L = 401 \text{ volts}$$

And:

$$E_C = 1.34 \times 100$$

$$E_C = 134 \text{ volts}$$

It is interesting to note also that at resonance a special condition in respect to the power factor occurs. Since the reactive elements cancel each other, R is equal to Z, the phase angle is 0°, apparent power is equal to true power, and the power factor is equal to 1. However, because of resonance, the current in the circuit is maximum and, accordingly, the voltages across the reactive elements may be very high, a condition that is characteristic of a circuit with a low power factor.

Example 6

What is the Z, TP, AP, $\angle\theta$, and power factor of a series RCL circuit shown in Fig. 9-15?

1. What is given?

   $E_A = 120$ V　　$X_L = 100$ ohms
   f = 60 Hz　　$X_C = 50$ ohms
   $R = 25$ ohms

2. What do you need to find?
   a. $Z = ?$
   b. $TP = ?$
   c. $AP = ?$
   d. $\angle\theta = ?$
   e. $PF = ?$

3. Find Z by using $Z=\sqrt{R^2+(X_L-X_C)^2}$, then substitute.

$$Z=\sqrt{25^2+50^2}$$
$$Z=\sqrt{625+2{,}500}$$
$$Z=\sqrt{3{,}125}$$
$$Z=55.90169944 \text{ ohms}$$

4. Find I_T by using $I_T=\frac{E_A}{Z}$, then substitute.

$$I_T=\frac{120}{55.90168844}$$
$$I_T=2.146625258 \text{ amperes}$$

5. Find AP:

$$\text{AP}=E\times I$$
$$\text{AP}=120\times 2.146625258$$
$$\text{AP}=257.595031 \text{ volt-amperes}$$

6. Find $\angle\theta$ by using $\cos\angle\theta=\frac{R}{Z}$, then substitute:

$$\text{Cos }\angle\theta=\frac{25}{55.90169944}$$
$$\text{Cos }\angle\theta=0.4472135955$$
$$\angle\theta=63.43494882°$$

7. PF = Cos $\angle\theta$. Since cos $\angle\theta=0.4472135955$, this is also the power factor. There is another way to find the power factor if the angle isn't known, but the TP and AP are known. PF is found by: $\text{PF}=\frac{\text{TP}}{\text{AP}}$.

8. To find TP you use $\text{TP}=E\times I\times\cos\angle\theta$.

$$\text{TP}=120\times 2.146625258\times 0.4472135955$$
$$\text{TP}=115.2 \text{ watts}$$

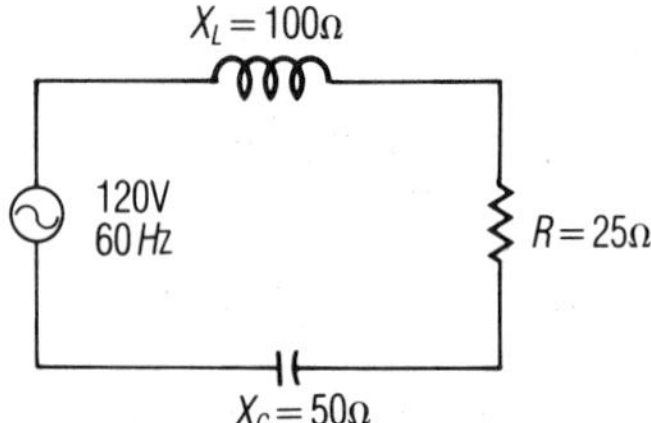

Fig. 9-15

Problems

Find the values needed to complete the circuit for a series RCL combination.

	E_A	I_T	X_L	X_C	R	AP	TP	$\angle\theta$	Z
a.	100V		25	50	25				
b.	50V		50	25	100				
c.	25V		10	5	10				

CHAPTER 10

Special-Purpose Circuits

TRANSFORMERS

The operation of a transformer depends on the principle of electromagnetic induction. Basically, a transformer consists of any two inductors (in separate circuits) so placed physically that the changing electromagnetic field set up by an alternating current in one induces an alternating emf in the other. Thus, mutual inductance exists between the coils, and the two circuits are said to be inductively coupled. When a basic transformer is connected between an a.c. generator and a resistance load, the coil connected to the source of power is called the *primary* winding and the coil connected to the load is called the *secondary* winding. See Fig. 10-1. The power delivered by the generator passes through the transformer and is delivered to the load, although no direct connection exists between the primary and the secondary winding, or between generator and load. The connection that does exist is the flux linkage between the coils, and power is effectively transferred by induction. Thus, the power consumed by the primary is equal to the power delivered by the secondary. If the coils of a transformer were completely shielded from each other, no power transfer could take place and the transformer would be useless.

For a maximum transfer of power from the primary to the secondary of the transformer the *flux linkage* must be complete. That means all the lines of force set up by the primary winding must link the secondary winding. That is why the secondary is often wound directly on the primary with only protective insulation separating the two coils. Then, since the reluctance (in magnetism, reluctance is resistance or opposition) of air is very great and its

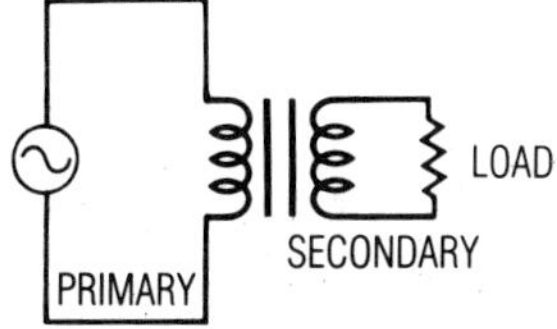

Fig. 10-1

permeability small, the introduction of a soft-steel core of high permeability in the transformer increases the flux linkage between the coils and makes possible a high percentage of power transfer. Even with the use of high-permeability cores, a few of the flux lines fail to link the secondary winding and are effectively lost. This constitutes a flux leakage which prevents the transformer from being a perfect conductor of power from the generator to the load. However, a well-designed iron-core transformer may effect a 98% flux linkage, which means that the *K*, or coefficient of coupling, between the coils is 0.98. Fig. 10-2 shows a typical iron-core transformer with the flux lines set up by the primary linking the secondary by means of the low-reluctance path of the core. The small leakages are also shown. A of Fig. 10-3 shows the shell type of transformer core, which is the most efficient core type, and B is a cross section of the windings as they usually appear. Each layer of wire is separated from the other by sheets of waxed paper, and the primary

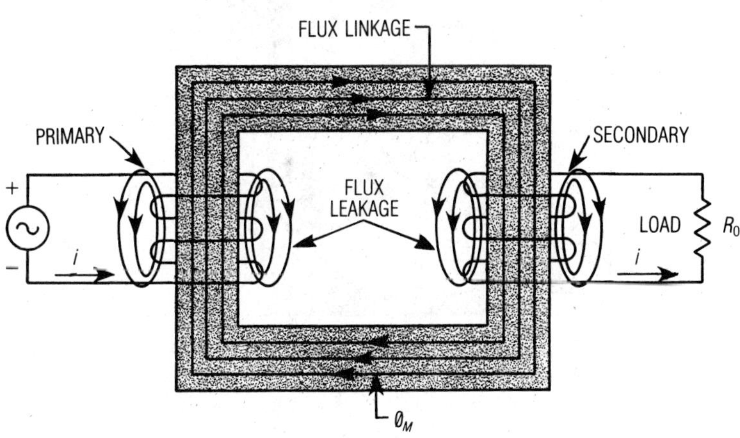

Fig. 10-2

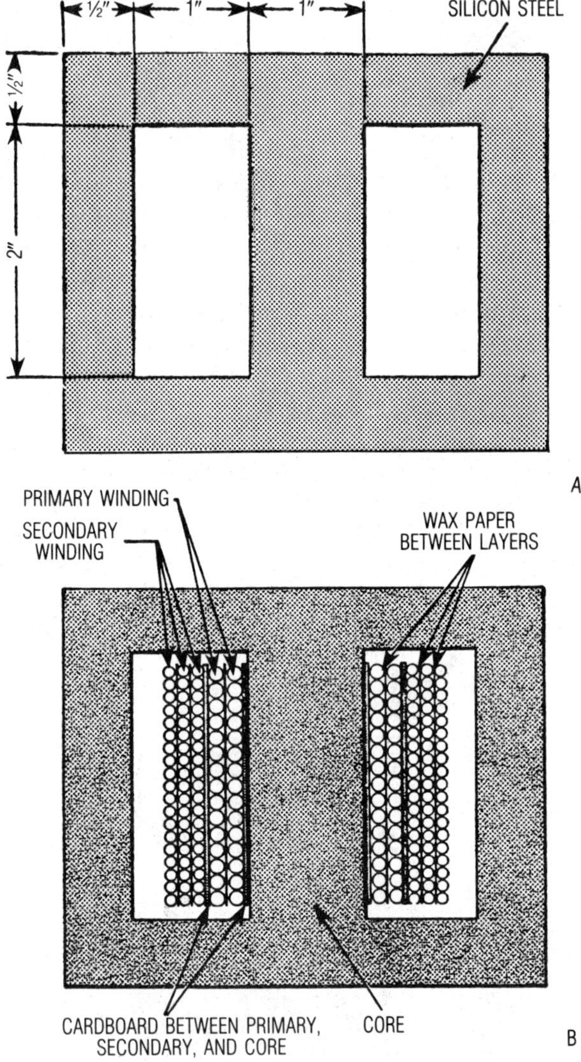
½"
1"
1"
SILICON STEEL
½"
2"
A
PRIMARY WINDING
SECONDARY WINDING
WAX PAPER BETWEEN LAYERS
CARDBOARD BETWEEN PRIMARY, SECONDARY, AND CORE
CORE
B

Fig. 10-3

winding is separated from the secondary winding by varnished paper, cardboard, or simply by the Formvar® insulation on the wire.

Theory of Operation

The ability of a transformer to transfer energy from its primary to its secondary through flux linkage is a function of inductive coupling or high mutual inductance. This means that the inductance of each winding should be as great as possible. If the transformer were ideal (the windings showing infinite inductance), the inductive reactance of the primary would be infinite and for any a.c. frequency. At the usual 60-Hz power line frequency, however, the inductance of the primary must be large in order to generate appreciative reactance (X_L). Thus, if the iron core of the 60-Hz transformer were removed, the inductive reactance would fall and the primary circuit would show a high current even with no load on the secondary. This initial magnetizing current in the primary of a transformer under no load should be kept as low as possible since it represents a loss. The greater the inductance, the greater the reactance and the less magnetizing current needed to set up the flux linkage.

Turns Ratio and Voltage

The magnitude of the voltage induced in either side of a transformer depends directly on the number of turns. Thus, the back emf induced by the changing current in the primary is not equal to the emf induced in the secondary *unless* the number of turns in the primary equals the number in the secondary. Furthermore, since the back emf in the primary is equal to the applied voltage, a ratio may be set up to determine the emf induced in the secondary in terms of the applied voltage and the turns ratio of the two coils. The formula for that would be:

$$\frac{E_P}{N_P} = \frac{E_S}{N_S}$$

E_P is the voltage applied to the primary
E_S is the voltage induced in the secondary
N_P is the number of turns in the primary
N_S is the number of turns in the secondary

The equation may also be written as:

$$E_P N_S = E_S N_P$$

Or:

$$E_S = \frac{N_P N_S}{N_P}$$

The expression $\frac{N_S}{N_P}$, or $\frac{N_P}{N_S}$, is called the *turns ratio* and may be expressed as a single factor.

Example 1

A transformer with 1,000 turns in the secondary and 250 turns in the primary has a turns ratio of 4:1 or 4. See Fig. 10-4. If 120 volts a.c. is applied to the primary of this transformer, what is the voltage induced in the secondary?

1. Choose the proper formula:

$$E_S = \frac{E_P N_S}{N_P}$$

2. Then substitute:

$$\frac{120\text{ V} \times 1{,}000\text{T}}{250\text{T}} = \frac{120{,}000}{250}$$

3. $E_S = 480$ volts

Such a transformer is called a *step-up* transformer. If the ratio $\frac{N_S}{N_P}$ is less than 1, the secondary winding has fewer turns than the primary and the secondary voltage is less than the applied voltage. Such a transformer is called a *step-down* transformer. It should be noted, however, that the terms *step-up* and *step-down* as applied to transformers always refer to the *voltage level*, not to the current level, nor are they ever applied to the power rating.

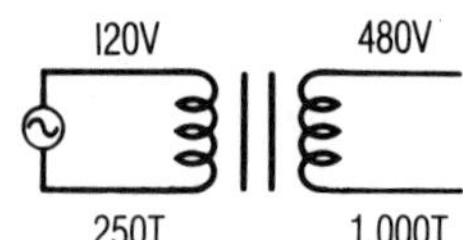

Fig. 10-4

Example 2
The voltage *ratio* of secondary to primary may be found by determining the number of volts per turn for a given transformer. Thus, 120 volts applied to a primary of 250 turns results in a volts-per-turn ratio of 0.48, or:

$$\text{Volts/turn} = \frac{120}{250} = 0.48$$

Since the number of volts per turn is a constant for any given transformer, the voltage in the secondary winding may be determined by multiplying this constant by the number of turns in the secondary, or:

$$E_S = 1{,}000 \times 0.48$$
$$E_S = 480 \text{ volts}$$

The *volts-per-turn ratio* is also convenient for determining a number of secondary voltages when a transformer carries more than one secondary winding.

Example 3
Figure 10-5 shows the schematic of a transformer with the primary winding linking a number of secondary windings. The primary consists of 200 turns, secondary winding S_1 is 1,200 turns, S_2 is 850 turns, S_3 is 11 turns, and S_4 is 22 turns. The volts per turn of this transformer is $\frac{E_P}{N_P}$, or $\frac{120}{200}$, and is equal to 0.6. Winding S_1, then, has an induced voltage of:

$$E_{S1} = 1{,}200 \times 0.6 = 720 \text{ volts}$$
$$E_{S2} = 850 \times 0.6 = 510 \text{ volts}$$
$$E_{S3} = 11 \times 0.6 = 6.6 \text{ volts}$$
$$E_{S4} = 22 \times 0.6 = 13.2 \text{ volts}$$

Turns Ratio and Current

In the transfer of electrical power across an ideal transformer, the power absorbed by the primary winding is equal to the power delivered by the secondary winding. That means $P_{pri} = P_{sec}$.

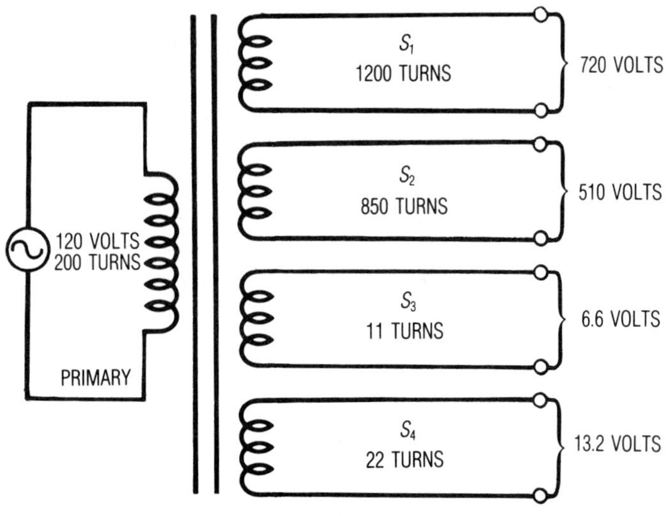

Fig. 10-5

In such an ideal transfer of energy from generator to load, the load appears as a pure resistance to the generator and apparent power is equal to true power. That means the power factor of the transformer is 1 and there is no phase angle. That also means the power in (primary) is equal to the power out (secondary). Formulas that can be used to state this follow:

$$P_P = E_P I_P$$

$$P_S = E_S I_S$$

$$E_P I_P = E_S I_S$$

$$\frac{I_P}{I_S} = \frac{E_S}{E_P}$$

However, the ratio of secondary voltage to primary voltage is equal to the turns ratio:

$$\frac{E_S}{E_P} = \frac{N_S}{N_P}$$

Therefore, the ratio of current in the primary to the current in the secondary is equal to the turns ratio:

$$\frac{I_P}{I_S}=\frac{N_S}{N_P}$$

Or:

$$I_P=I_S\left(\frac{N_S}{N_P}\right)$$

Example 4

Current across a transformer varies *inversely* as the number of turns. In Fig. 10-6 a simple power transformer of 300 turns on the primary and 900 turns on the secondary is connected to a 110-volt a.c. line and a 165-ohm load. The turns ratio is 3 to 1, and E_S is equal to 330 volts. Then, since the load determines the energy used, what is the current in the secondary?

1. Select the appropriate formula:

$$I_S=\frac{E_S}{R_O}$$

2. Then substitute:

$$I_S=\frac{330}{165}$$
$$I_S=2 \text{ amperes}$$

3. The current in the primary is found by:

$$I_P=I_S\left(\frac{N_S}{N_P}\right)$$

4. Then substitute:

$$I_P=2\times 3$$
$$I_P=6 \text{ amperes}$$

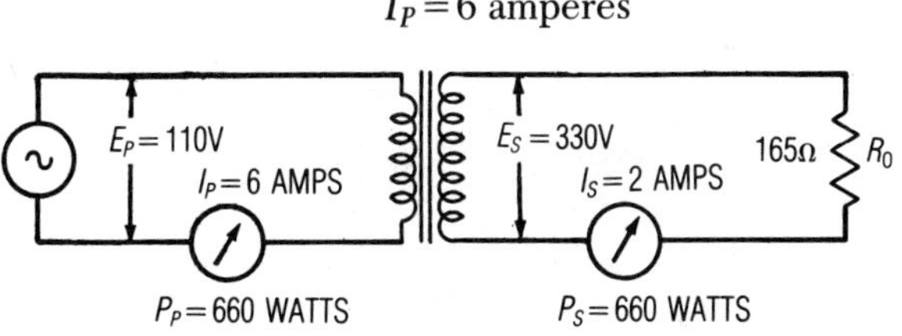

Fig. 10-6

A turns ratio of 3 to 1 increases the applied voltage from 110 to 330 volts and at the same time decreases the current from 6 amperes in the primary to 2 amperes in the secondary. From these observations, you can see that the product of current and voltage in one side of a transformer is equal to the product of current and voltage in the other side. Thus, the power in the primary circuit is 110 times 6, or 660 watts. The power consumed by the secondary circuit is 330 times 2, or 660 watts.

Transformer Efficiency

So far, the ideal transformer has been assumed. Such an ideal transformer has 100% efficiency—that is, the ratio of power output compared to power input is 1. Practical iron-core transformers are not 100% efficient, but when carefully designed they show a high efficiency ranging from 95 to 99%. This high efficiency is possible in a transformer because of the careful attention devoted to minimizing the effective losses due to flux leakage, hysteresis, eddy currents, flux saturation of the core, as well as the copper losses of the coil windings and losses from distributed capacitance.

Hysteresis

The mutual inductance between windings of a transformer depends directly on the coefficient of coupling between them. Total flux linkage is impossible. But by interleaving windings—that is, by winding first a layer of one coil and then a layer of the other—and by the use of cores of high permeability, flux leakage is held to a minimum.

The effort to achieve a high coefficient of coupling at power line frequencies (60 Hz in the United States and 50 Hz in Europe) requires cores of high permeability but involves in turn the loss from *hysteresis*, which is the lagging of the magnetization and demagnetization of the soft-steel core behind the alternating current in the circuit. The hysteresis loop is a graphical representation of this phenomenon, plotting flux density at any instant against the ampere-turns causing it. See Fig. 10-7. Thus, the flux density of a given core material is at point D (well above zero) when H, the force which originally had caused this density, is at zero. The flux density does not reach zero until point E, when the magnetizing

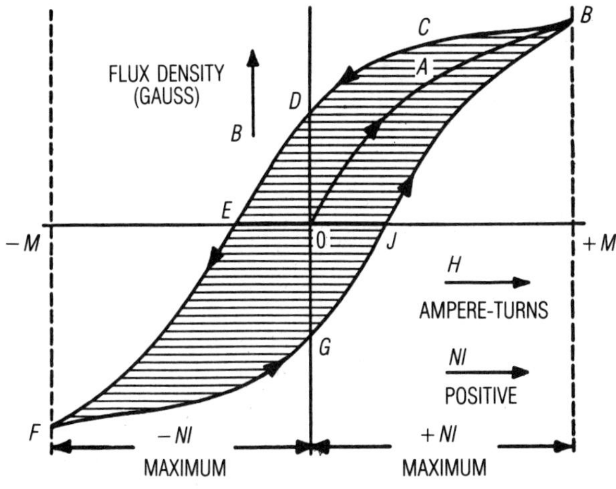

Fig. 10-7

force is in the opposite direction. On the return swing, the flux density is at point *G* when *H* is zero, and zero when *H* is at point *J*. This persistent lagging of the magnetization behind the current causing it means that if a sine wave of current is applied to an iron core transformer, the resultant flux density curve will not be sinusoidal and the maximum effect of the current will not be realized. Figure 10-8 shows a core specimen, its hysteresis loop, the sine wave of current causing it, and the resultant flux density curve. Since the width of the hysteresis loop horizontally at the *x*-axis and vertically at the *y*-axis is a sign of the amount of lag of the magnetization behind the current, the area of the hysteresis loop is proportional to the loss. This loss can be lessened by careful selection of a core material such as soft silicon steel.

Saturation

In addition to hysteresis losses, soft-steel cores show losses due to saturation. That is, the number of flux lines in the core reaches a point at which an increase in current causes no additional magnetization or less additional magnetization than the increase in magnetizing force warrants.

Figure 10-9 shows a saturation curve, which is a graph of flux density plotted against a steadily increasing direct current. At the

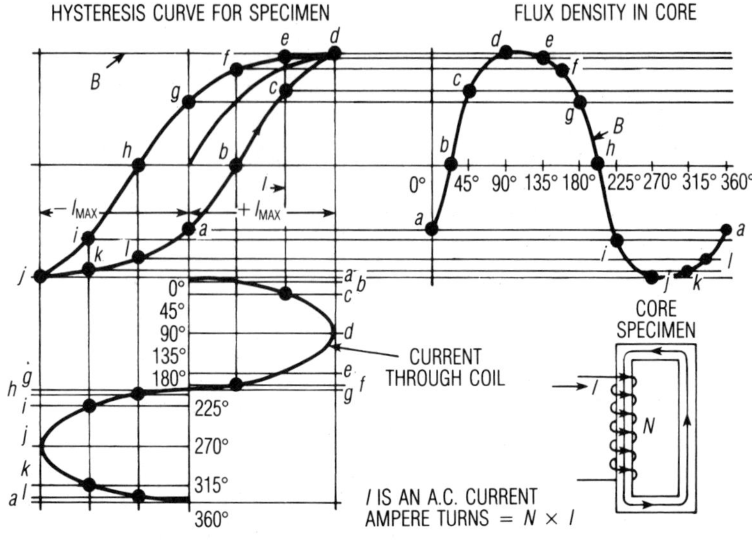

Fig. 10-8

knee of the curve, the linear increase in flux density ceases and there is little increase in flux density for a large increase in current.

Thus, for a rise in current from 0 to A, the flux density rises from A to C, but for an equal rise in current from B to D, the flux density rises only from E to F. Hence, increasing current beyond the knee of the saturation curve represents an increasing loss in

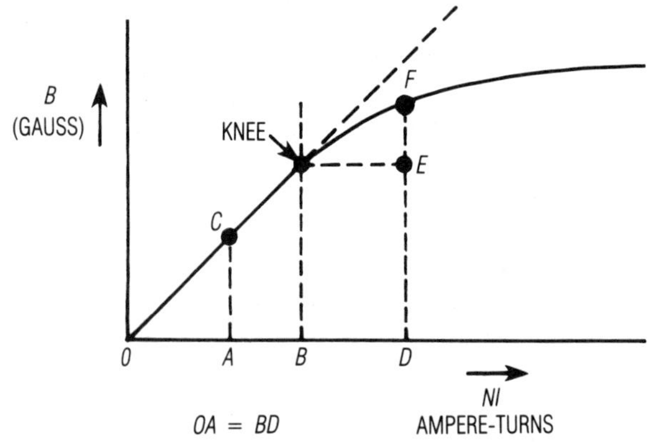

Fig. 10-9

efficiency since the same effect of magnetization could be obtained with a much smaller current were the core not saturated. An increase in cross-sectional core area decreases the flux density for a given current, but increases weight, cost, and hysteresis loss of the transformer. For this reason, most transformers designed for power transfer at 60 Hz operate at or near the knee of the saturation curve of the core material.

The permeability of a magnetic core falls off when flux density increases beyond an optimum point. This is equivalent to saying that as the core saturates, the ratio of B to H (flux density to magnetizing force), which is the definition of permeability, becomes smaller. In addition, a certain amount of direct current will be flowing in an a.c. circuit where certain circuits are involved. This direct current magnetizes the core permanently in one direction and decreases the magnetizing variation possible for the alternating current present. Because of this fact, iron-core chokes (except swinging chokes or saturable reactors) and iron-core transformers are designed so that a small *air gap* exists in the magnetic circuit.

The reluctance of the air gap is high compared to the reluctance of the soft-steel core, and the overall reluctance of the circuit is increased, reducing inductance but effectively preventing saturation. Physically, the air gap needed is very small and may be achieved by the end-to-end placing of the laminations or by inserting a thin piece of plastic or paper through the core at this point. Many additional turns of wire are needed when an air gap is used in order to achieve the same inductance. But, more important, the choke or the transformer action, which depends on flux change in proportion to changing current, is not lost through saturation of the core.

Eddy Currents

Magnetic core materials cause additional transformer losses because, as conductors, small short-circuited currents, called eddy currents, are induced in them. To reduce eddy currents to a minimum, transformer cores are laminated—that is, made up of thin strips of steel pressed together. Each strip is sprayed with an insulating coating so that the d.c. resistance between strips is very high. Thus, eddy currents find a series of high-resistance paths, whereas the magnetic permittivity of the core is not affected.

Figure 10-10 shows in A the eddy currents in a cross section

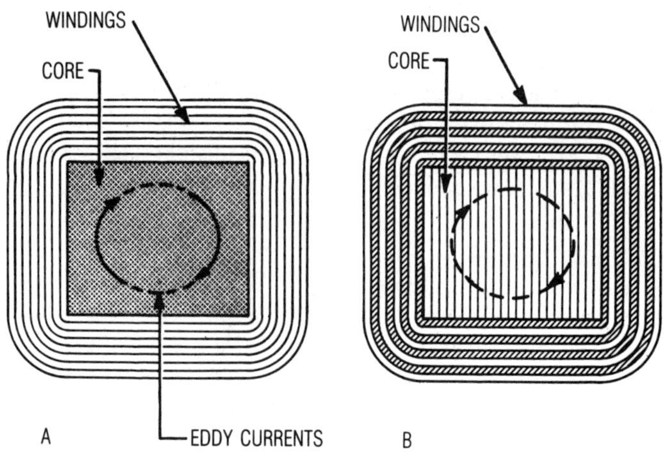

Fig. 10-10

of unlaminated core; in B the same cross section is shown with a laminated core.

Copper Losses

The current in the primary and secondary windings of a transformer must flow through the d.c. resistance of the wire. A certain amount of power is lost through heat. Heat loss is true power (I^2R) loss. However, transformers carrying considerable amounts of power use wire of large cross-sectional area. This cuts down on heat loss. In addition, it lowers the permeability of the core and increases resistance, thus increasing other losses. However, since a high percentage of flux linkage at power line frequencies requires large inductance, some compromise between the size of the core and the number of windings must be made. Thus, a huge core and small winding for a given inductance would have little copper loss, but the transformer would be heavy and awkward and core losses would be great. On the other hand, a small core reduces core loss and increases copper losses because of the increased number of turns. But for most applications, the d.c. resistance of the wire may be ignored if the ratio of inductive reactance to resistance is 10 to 1 or greater. Skin effect, or the tendency of alternating currents to travel on the surface of a conductor, also raises the resistance of the wire, but this effect, although present at all frequencies, is more noticeably a loss factor at the higher frequencies. Distributed capacitance, the stray capacitance existing between closely packed windings, is also a loss factor but more noticeable at the audio and radio frequencies than at the power frequencies.

Power Transformers

Transmitting large amounts of power over long distances means the power consumed by the resistance of the conducting wires is lost as heat. This true power loss varies directly as the resistance and the square of the current since it is equal to I^2R. If current in the line is doubled, the power lost is four times as much, the loss varying as the square of the current. Or, if the current in the line is halved, the power lost is only one-quarter as much. Thus, power generated at Niagara Falls may be sent to New York City, or generated in the Tennessee Valley and sent to Atlanta, with little loss if the current is kept as low as possible. This is done by means of a power transformer. This type of transformer converts the power output of the generators to a high-voltage, low-current level for transmission and then reconverts this power to a low-voltage, high-curent level for consumption in the home.

For instance, if 1,000 watts of power (5 amperes at 200 volts) is sent through 10 ohms of line, the power loss is I^2R (5^2 times 10), or 250 watts. Converting this power to 0.5 amperes at 2,000 volts, the loss is 0.5^2 times 10, or 2.5 watts, a saving of 247.5 watts, and only a 1% loss.

Long-distance power transmission uses voltage levels as high as 750,000 volts with very small currents. The transformers used to convert power at this level are huge units filled with oil to prevent arcing-over within the windings because of the high voltage. An electric substation is a series of such transformer units for reconverting power, and the familiar neighborhood pole transformer brings voltage and current to the levels used in the home, 110 to 120 volts at the current demanded.

The power transformers used in radio and communications are much smaller units than the power transformers, but their function is the same. The efficient transfer of power while changing the voltage-current level is their primary function. A transformer made for use on power line frequencies should be operated within 90% of its rated load since the current in the secondary brings voltage and current in phase, reduces the phase angle, and brings the power factor close to unity.

Any power transformer operated at one-fifth its proper load shows considerable primary and secondary inductive reactance. The induced voltages across it are very high. This causes arcing and breakdown of insulation with the resultant burning out of the transformer. Under full load, the inductive reactance of the primary circuit is almost completely canceled by the opposing magnetizing force set up by the secondary current, the phase angle is small and, accordingly, the power factor is close to unity.

Any reactance present in the transformer is a *leakage* reactance. This leakage affects the efficiency of the transformer. A carefully designed power transformer has a very small leakage reactance and small copper and core losses. This means it shows a high power factor and high efficiency. Commercial power transformers generally show a power factor of 0.9 or above and an efficiency of approximately 95%.

Autotransformers

The autotransformer is a special type of power transformer that is designed for good voltage regulation under varying loads. A single tapped winding characterizes the autotransformer. See Fig. 10-11. When used as a step-up transformer, as in A, all the primary winding is part of the secondary winding; and when used as a step-down transformer, all of the secondary winding is part of the primary winding, as in B. Voltages across the individual windings follow the turns ratio. But it must be remembered that certain turns are common to both coils, and that generally the coils are tapped for variation of the turns ratio, with resultant change in voltage levels. A disadvantage of this arrangement is that the secondary current is not electrically isolated from the primary circuit as it is in a transformer in which the only connection between the source and the load is the magnetic coupling of the coils.

The chief advantage of the autotransformer is that the load may be varied without arcing and with little change in output voltage.

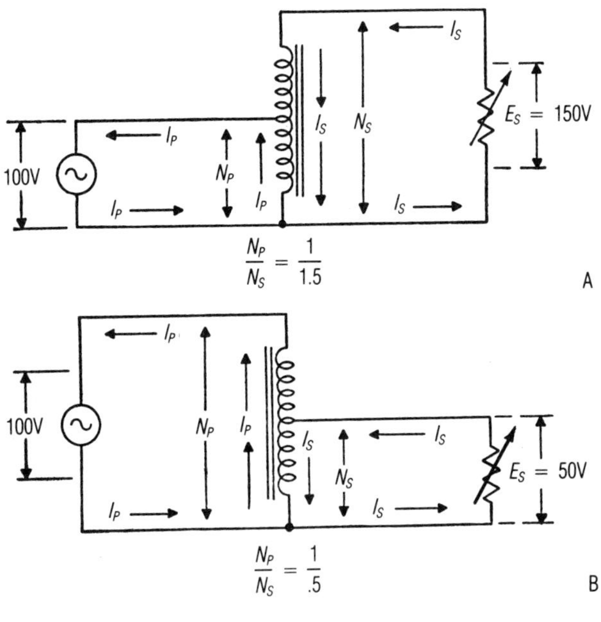

Fig. 10-11

This result depends on the fact that primary and secondary currents in any transformer are 180° out of phase and therefore tend to cancel in that part of the transformer which is common to both windings. For example, A of Fig. 10-11 shows an autotransformer with a ratio of 1 to 1.5. With 100 volts applied to the primary and a 75-ohm load on the secondary, E_S is 150 volts and I_S is 2 amperes. By the turns ratio formula, I_P should be 3 amperes. However, since I_S and I_P are 180° out of phase, the total current in the primary is the algebraic sum of the two, or 1 ampere. If the load on the secondary is increased to 4 amperes, I_P increases to 6 minus 4, or 2 amperes. If the load on the secondary is decreased to 1 ampere, I_P decreases to 1.5 minus 1, or 0.5 ampere. B of Fig. 10-11 shows this same relationship in terms of a step-down autotransformer with a turns ratio of 1 to 0.5. With the same voltage applied and a load of 25 ohms, E_S is 50 volts and I_S is 2 amperes. The current in the part of the primary common to both is 1 ampere in the direction of the secondary current. Thus, the primary current of 1 ampere may be conceived as flowing directly to the load and not passing through the primary winding. Any variation in the load causes a smaller change in primary current than is possible with the separate winding power transformer.

The further practical effect of these small primary currents for a given load on the secondary is that power loss in the autotransformer is also small. For example, in A of Figure 10-11 the 1-ohm d.c. resistance of the primary winding results in an I^2R loss of 1 watt, 4 watts, and 0.25 watt for the three loads listed, as compared to losses of 9, 36, and 2.5 watts in a separate primary. The I^2R loss in the secondary is computed only for that part of the secondary not common to the primary. If you have a decrease in voltage, in both primary and secondary, caused by the d.c. resistance of the windings under heavy load, the loss is much less in an autotransformer. Therefore, these tapped transformers are widely used to vary line voltage for a given load, and to maintain a given output voltage under varying load conditions.

PROBLEMS

What is the secondary voltage and the primary current for the following transformers?

E_P (volts)	E_S (volts)	N_P (turns)	N_S (turns)	I_P (amperes)	I_S (amperes)
120		100	300		0.333333333
240		300	900		0.333333333

DELTA AND WYE THREE-PHASE POWER CIRCUITS

Most commercial sources of power in the United States produce three-phase (3ϕ) electricity. What we have studied so far, and what you have in your home, is single-phase (1ϕ) electricity. This single-phase electricity is obtained from the transmission lines by using three single-phase transformers connected to the three-phase line.

It is sometimes convenient and necessary to generate two or three separate voltages in the same generator. These voltages are of the same amplitude and frequency. However, they are out of phase with one another. These out-of-phase voltages are referred to as *polyphase*. Each leg of a polyphase circuit is called a *phase*. *Poly* means "more than one," so it is apropos for either two-phase or three-phase a.c. sources. Figure 10-12 shows a two-phase generator. The two armature windings are wound on the stator and placed 90° apart physically. The field magnet is the rotor that is excited by direct current supplied through slip rings. The two armature coils may be wired to separate terminals and used as individual a.c. supplies or they may be connected as in (b) of Fig. 10-12. Voltages that are generated by a two-phase generator are 90° out of phase. Take a look at Figure 10-12(c). This type of voltage aids in starting motors and makes for smoother operation of motors.

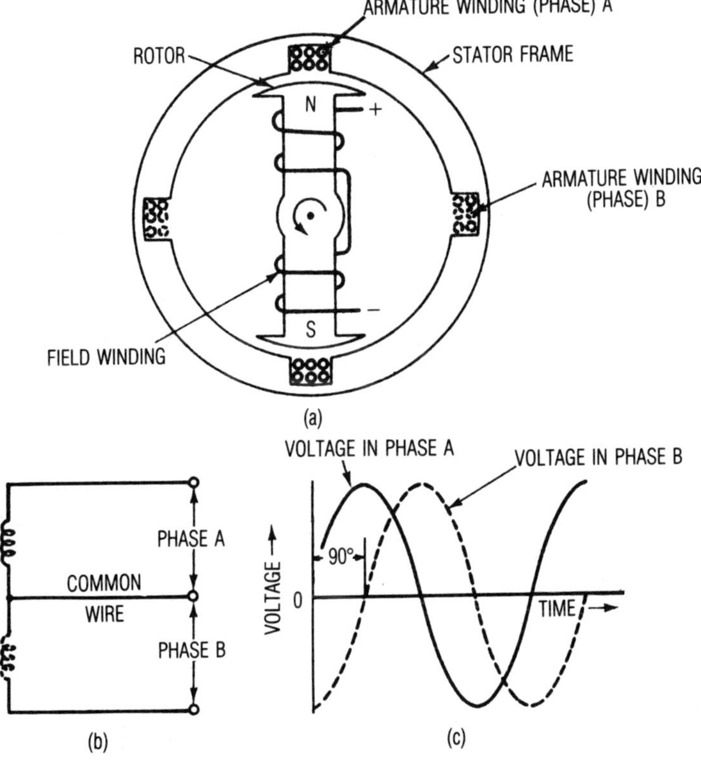

Fig. 10-12

However, it is not available in the United States. The three-phase is preferred because the motors do not need starting devices and can run without maintenance for years.

Nicola Tesla was the man most responsible for the use and generation of three-phase electricity. The three-phase generator is shown in Fig. 10-13. Note the three independent phases wound on the stator and placed 120° apart physically. Figure 10-13(b) shows how the three-phase power looks on a time basis. You can see how much smoother the operation of a 3ϕ motor is, due to the continuous supply of power by the three phases.

The two ways that three-phase power is connected are shown in Fig. 10-13(c). These two modes of connection are called *delta* and *wye*. The Greek letter *delta* is in the shape of a triangle (Δ). The wye connection gets its name from the letter *y*. Note how the

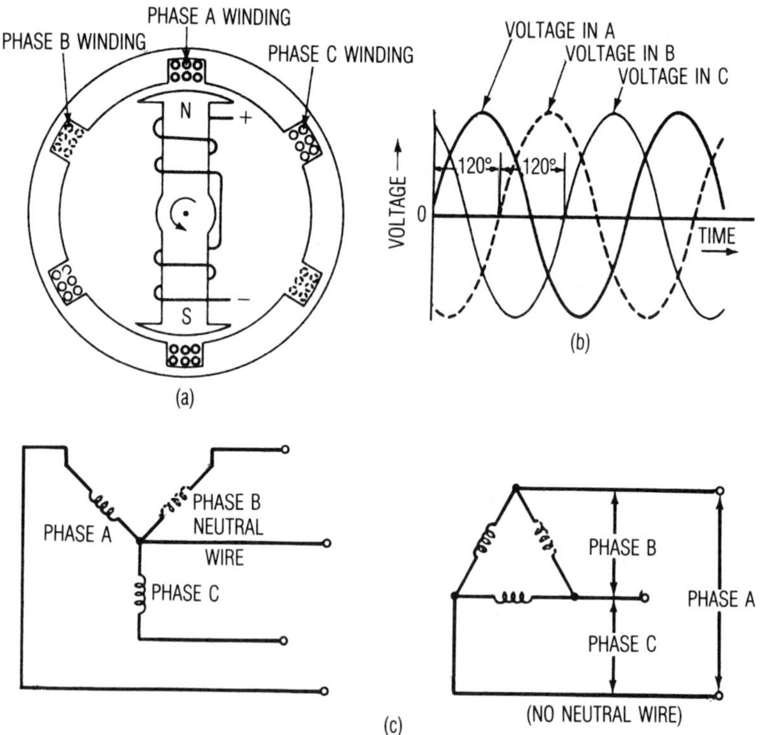

Fig. 10-13

delta connection does not have any place for a neutral wire. The wye connection has the ground or neutral wire.

The automobile generator is a three-phase alternator. It produces a steady output at low speeds and can furnish enough power for the taillights, headlights, radio, windshield wipers, and other accessories at idle speed, and while you are driving in traffic it can still trickle-charge the battery. The d.c. generator of older cars was unable to charge the battery and operate the electrical demands of the car while idling.

Three-Phase Connections

The three-phase connections can be: wye to wye, delta to delta, delta to wye, wye to delta. These connections are necessary to obtain the proper voltages and current for certain loads. The loads will also be discussed here and formulas for figuring the loads will be presented.

Delta and Wye Connections

In a balanced circuit, when the generators are connected in *delta*, the voltage between any two lines is equal to that of a single phase. The line voltage and the voltage across any winding are in phase, but the line current is 30° or 150° out of phase with the current in any of the other windings. See Fig. 10-14. In the delta-connected generator the line current from any one of the windings is found by multiplying the phase current by the square root of 3, or 1.73.

In the *wye* connection, the current in the line is in phase with the current in the winding. The voltage between any two lines is not equal to the voltage of a single phase, but is equal to the vector sum of the two windings between the lines. The current in line *A*

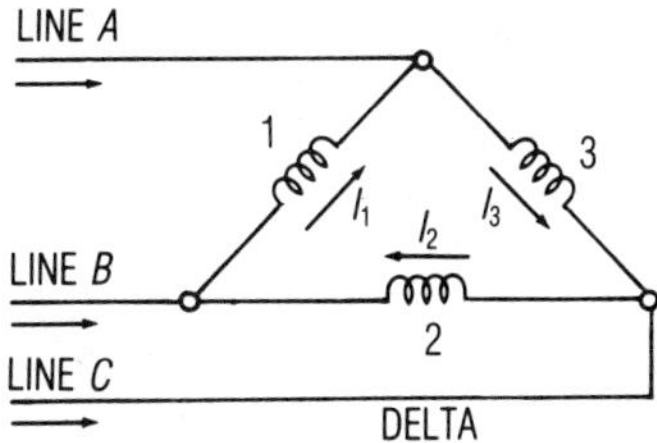

Fig. 10-14

of Fig. 10-15 is the current flowing through the winding L_1; that in line B is the current flowing through the winding L_2; and the current flowing in line C is that of the winding L_3. Therefore, the current in any line is in phase with the current in the winding that it feeds. Since the line voltage is the vector sum of the voltages across any two coils, the line voltage E_L and the voltage across the windings $E\phi$ are 30° out of phase. The line voltage may be found by multiplying the voltage of any winding $E\phi$ by 1.73.

Commercial three-phase voltage from power lines is usually 208 volts. The standard values of single-phase voltage can be supplied from the line as shown in Fig. 10-16. The windings represent wye-connected transformers. Figure 10-17 shows the types of connections available in three-phase power. Figure 10-18 shows the voltages available for local use from three-phase connections when single-phase is required from the three-phase line serving a subdivision.

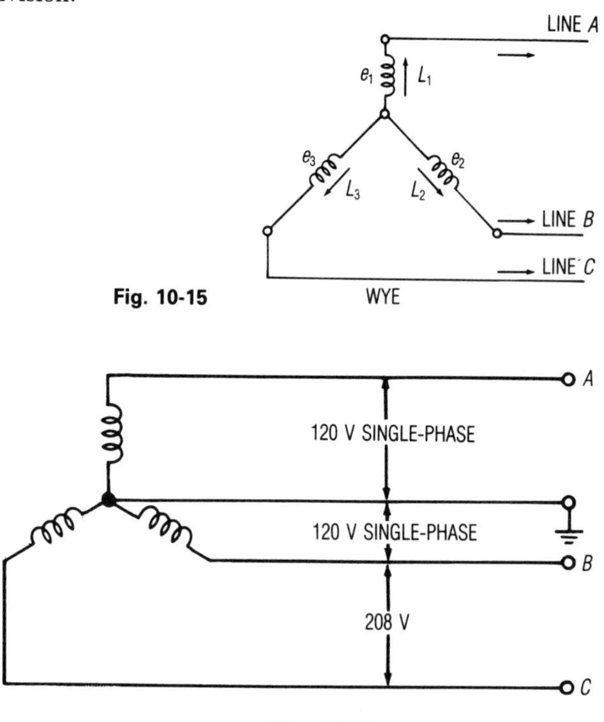

Fig. 10-15

Fig. 10-16

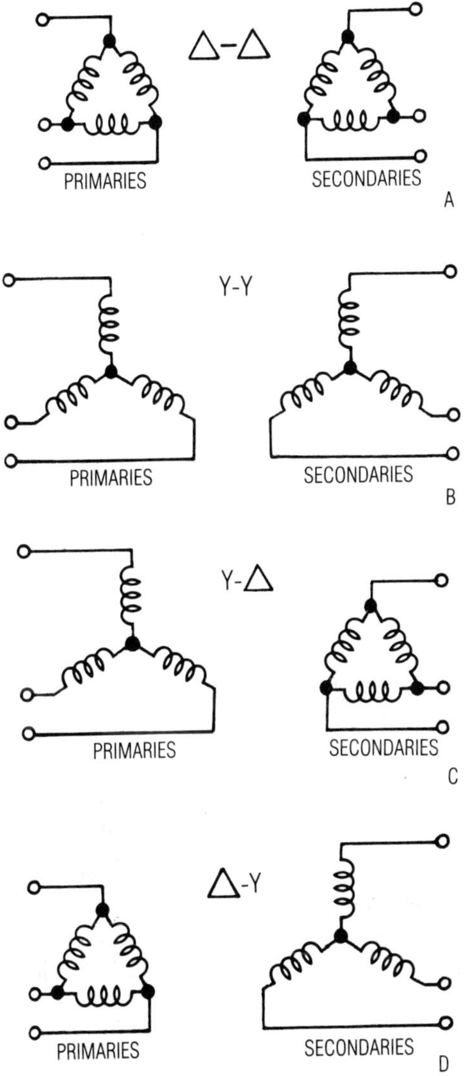
Δ-Δ
PRIMARIES
SECONDARIES
A
Y-Y
PRIMARIES
SECONDARIES
B
Y-Δ
PRIMARIES
SECONDARIES
C
Δ-Y
PRIMARIES
SECONDARIES
D

Fig. 10-17

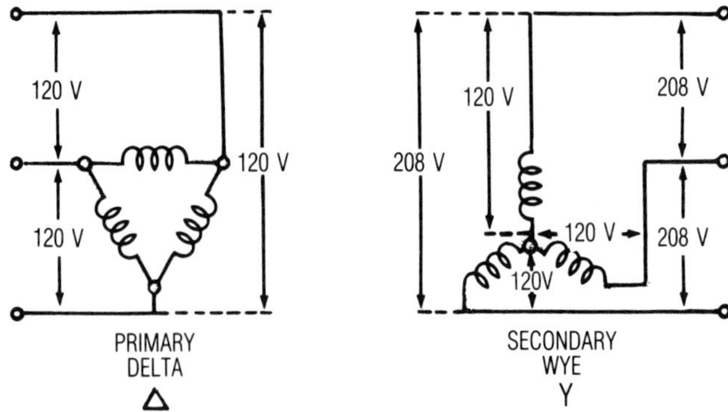

Fig. 10-18

PROBLEMS

1. What is the voltage available from a single phase of a three-phase wye-connected transformer, such as in Fig. 10-18?
2. What is the available voltage from a wye-connected transformer when two phases are tapped, as in Fig. 10-18?
3. If the voltage output from a wye-connected transformer is 120 volts for one winding, what is the output from the transformer when two windings are utilized?
4. What is the advantage of using a delta-to-delta-connected transformer?
5. Which type of transformer (delta, wye) has an advantage when it comes to connecting for current?
6. Which type of transformer (delta, wye) has an advantage when it comes to connecting for voltage?
7. What is the output in volts when two phases of a three-phase delta secondary are connected and each winding produces 120 volts?
8. When your home is supplied with 240 volts, which type of winding do you think is used to produce the 240 volts?
9. If your home uses 120-volt single-phase power, where do you think it comes from, a delta or wye source?

10. Which of the three-phase connections provides for a ground or neutral wire?

THREE-PHASE CIRCUITS

The three-phase circuits are usually referred to as delta and wye. They are most commonplace in three-phase power discussions and utilization.

Delta and wye connections are used in connecting transformers and in connecting the loads to these power sources. A three-phase delta resembles the Greek letter *delta* (Δ) in shape, and the three-phase wye is shaped like the letter *Y*.

As can be seen, the wye connection in Fig. 10-19 has terminals labeled *a*, *b*, and *c* and a common point shown at 0. The terminal pairs, that is, *a–b*, *b–c*, *c–a*, provide the three-phase supply. In this connection the line voltage is $\sqrt{3}$ (or 1.732050808) times the coil voltage, while *the line current is the same as the coil current*. The neutral point normally is grounded. It can be brought out to the power-consuming device by means of a four-wire power system for a dual voltage supply. If three wires are used and connected to points *a*, *b*, and *c*, then three-phase power is available for use with motors and other loads. If you are using a 208-volt system, you know that you have a wye-connected transformer supplying the power. Single-phase alternating current is available if you connect from 0 to any one of the points *a*, *b*, or *c*. From 0 to *a* will produce 120 volts of single-phase a.c., from 0 to *c* will produce 120 volts of

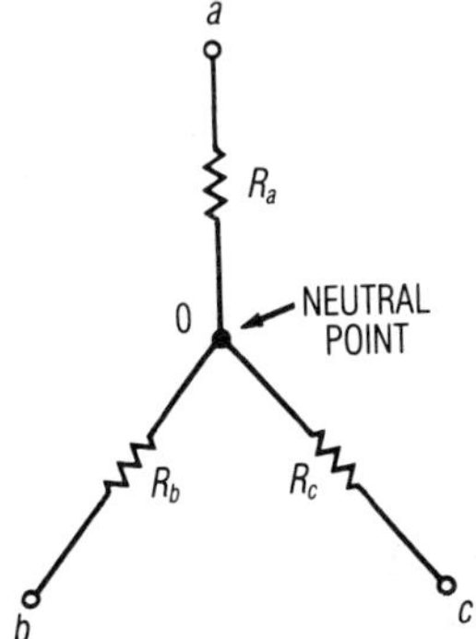

Fig. 10-19

single phase a.c., and from 0 to b will produce 120 volts a.c., single-phase. Single-phase a.c. is available, then, since $120 \times 1.732050808 = 207.846097$ volts. The common designation is 208 volts.

You probably have noticed that some fluorescent lamps in offices and factories use 208 volts. You may also have noticed that some larger electric heating systems use 208 volts.

Keep in mind that the delta connection has an advantage in current production because two of the coils are in series with each other and these two are in parallel with the remaining coil. Parallel arrangements do provide more current. The voltage available from any two terminals is the same. This voltage is also single-phase. However, if you wish 240 volts of three-phase a.c., all you need to do is bring out three wires, one each from point a, point b, and point c. The current available in a three-phase delta connection is $\sqrt{3}$ times the current capacity of any one coil.

Most power generated by commercial generators is three-phase. This means the electrician must have at least a passing knowledge of this type of system. The next few paragraphs will show how the transformer is loaded.

Delta and Wye Resistor Circuits

A three-terminal resistor network can be connected as either a delta or a wye network. See Figs. 10-19 and 10-20 for the resistor networks in their simplest forms.

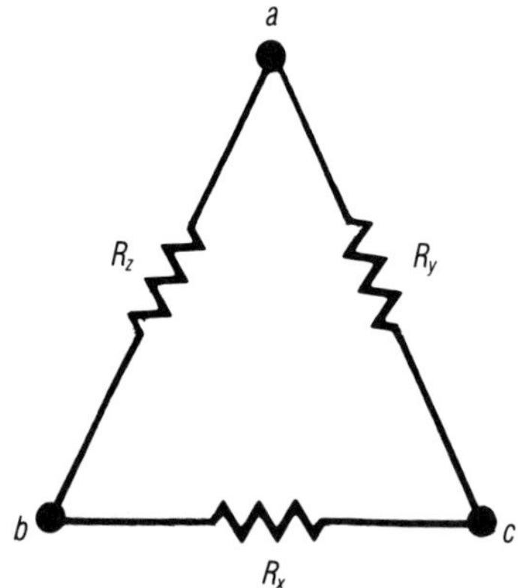

Fig. 10-20

In the wye network, the resistances between the terminals are easily determined:

$$a \text{ to } b = R_a + R_b$$
$$a \text{ to } c = R_a + R_c$$
$$b \text{ to } c = R_b + R_c$$

In the delta network, the resistances between the terminals are easily determined. The resistances between the terminals may be determined by combining the formulas for series and parallel resistance:

$$a \text{ to } b = \frac{R_z \cdot (R_x + R_y)}{R_z + (R_x + R_y)}$$

$$a \text{ to } c = \frac{R_y \cdot (R_x + R_z)}{R_y + (R_x + R_z)}$$

$$b \text{ to } c = \frac{R_x \cdot (R_z + R_y)}{R_x + (R_z + R_y)}$$

To simplify resistor networks for determining equivalent resistances, it is frequently necessary to convert a delta network to a wye, or a wye to a delta. The mathematics for such conversions are simple and are based on the formulas for series and parallel resistance.

To convert from *delta to wye*, the formulas to be used are:

$$R_a = \frac{R_y \times R_z}{R_x + R_y + R_z}$$

$$R_b = \frac{R_x \times R_z}{R_x + R_y + R_z}$$

$$R_c = \frac{R_x \times R_y}{R_x + R_y + R_z}$$

To convert from *wye to delta*, the formulas to be used are:

$$R_x = \frac{(R_a \cdot R_b) + (R_b \cdot R_c) + (R_c \cdot R_a)}{R_a}$$

$$R_y = \frac{(R_a \cdot R_b) + (R_b \cdot R_c) + (R_c \cdot R_a)}{R_b}$$

$$R_z = \frac{(R_a \cdot R_b) + (R_b \cdot R_c) + (R_c \cdot R_a)}{R_c}$$

Example 5

In the circuit shown in Fig. 10-21, *convert* the resistors connected in wye configuration to their equivalents in delta. (All resistors are in KΩ, so drop the 000 and add three zeros to the answer.)

$$R_x = \frac{(10 \cdot 20) + (20 \cdot 30) + (30 \cdot 10)}{10}$$

$$R_x = \frac{200 + 600 + 300}{10}$$

$$R_x = \frac{1,100}{10} = 110,000\ \Omega$$

Since the formulas for R_y and R_z use the same numerator as the formula for R_x, the solution for R_y and R_z are as follows:

$$R_y = \frac{1,100}{20} = 55,000\ \Omega$$

$$R_z = \frac{1,100}{30} = 36,666.67\ \Omega$$

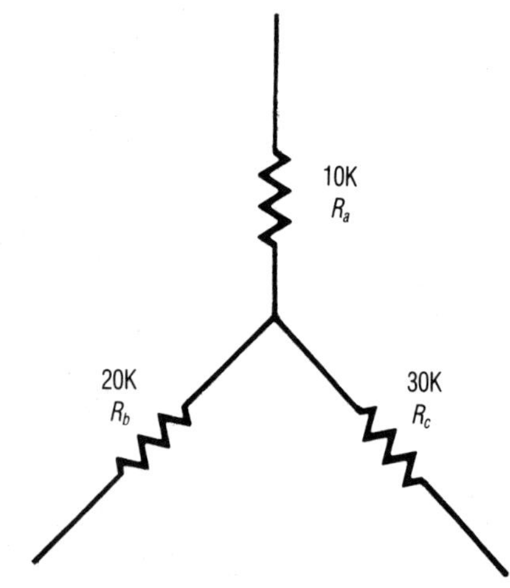

Fig. 10-21

Example 6
In the circuit shown in Fig. 10-22, *convert* the resistors connected in delta configuration to their equivalent in a wye. (Since all resistors are in thousands, leave off the three zeroes and add to the answer, or multiply the answer by one thousand.)

$$R_a = \frac{R_y \times R_z}{R_x + R_y + R_z}$$

$$R_a = \frac{20 \times 10}{30 + 20 + 10}$$

$$R_a = \frac{200}{60}$$

$$R_a = 3{,}333.33 \text{ ohms}$$

Since the formulas for R_b and R_c use the same denominator as the formula for R_a, the solution for R_b and R_c is as follows:

$$R_b = \frac{R_x \times R_z}{R_x + R_y + R_z}$$

$$R_b = \frac{30 \times 10}{60} = \frac{300}{60} = 5{,}000\ \Omega$$

$$R_c = \frac{R_x \times R_y}{R_x + R_y + R_z}$$

$$R_c = \frac{30 \times 20}{60} = \frac{600}{60} = 10{,}000\ \Omega$$

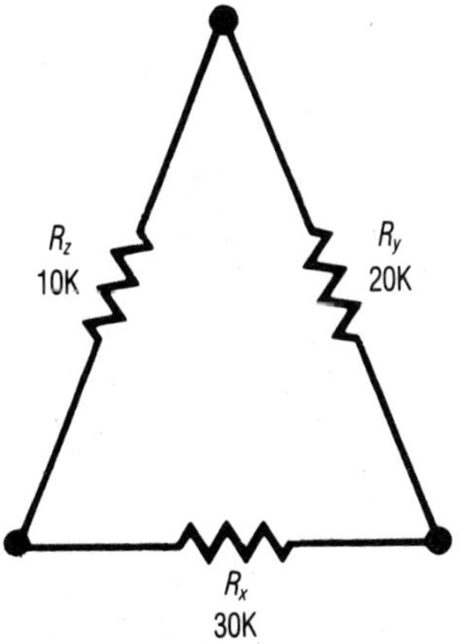

R_z
10K
R_y
20K
R_x
30K

Fig. 10-22

PROBLEMS

1. The values in Fig. 10-23 are given below. Find the missing values and complete the boxes by filling in the values. Convert Y to Δ.

R_a	R_b	R_c	R_x	R_y	R_z
1,000	5,000	10,000			

2. The values in Fig. 10-24 are given below. Find the missing values and complete the boxes by filling in the values. Convert Δ to Y.

R_x	R_y	R_z	R_a	R_b	R_c
5,000	10,000	30,000			

BRIDGE CIRCUITS

The Wheatstone bridge is used to get more accuracy in the measurement of resistance by mechanical devices. It uses four resistors connected as shown in Figure 10-25. The meter is used to detect any current flow. If the two branches of resistors (R_1 and R_3) and (R_2 and R_4) are balanced, there is no current flow through the meter.

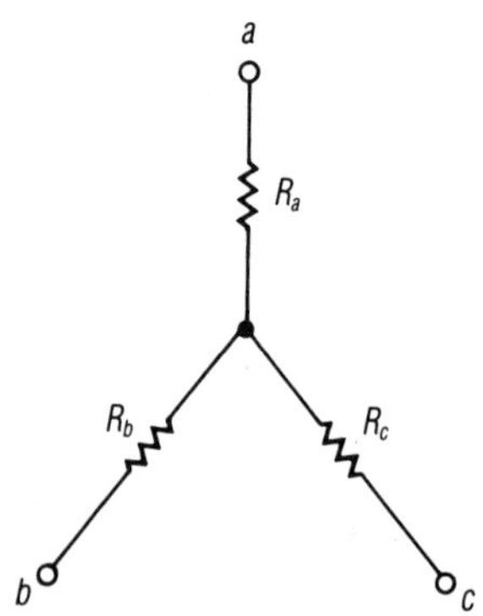

Fig. 10-23

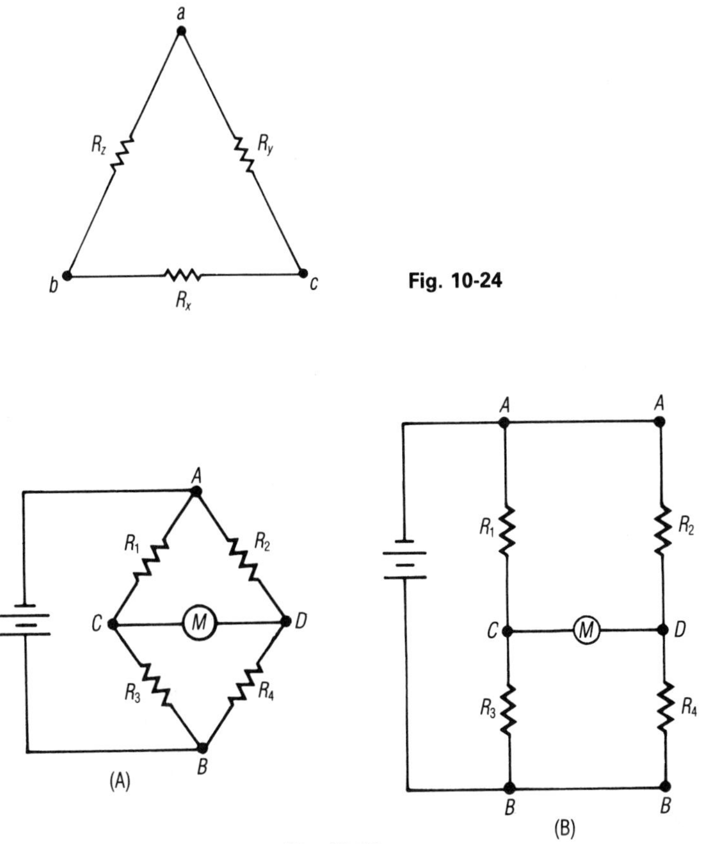

Fig. 10-24

Fig. 10-25

It is customary to design the bridge so that R_4 is the unknown resistor while R_3 is the variable one used to adjust until the current does not flow in the meter. R_3 is usually a resistance substitution box with resistors which vary in 1-ohm, 10-ohm, 100-ohm, or 1,000-ohm steps. R_1 and R_2 are fixed in value. The meter is usually a galvanometer that is sensitive to microamperes with a zero center. This way you can tell which way the resistor has to be adjusted to balance the bridge circuit.

The bridge is so connected that an unknown resistor is placed between points B and D. The circuit is energized and the meter is checked for any current flow through it. If there is current flow, the adjustable resistor, R_3, is adjusted until there is no current flow through the meter and it reads zero. This condition exists only when there is no potential difference between the two points of the meter (C and D). Let's take an example to show what is meant by the statement and also to see how this bridge can be used to check resistance.

Example 7

Assume that the applied voltage is 12 volts. R_1 is 1,000 ohms, R_2 is 400 ohms, R_3 is 2,000 ohms, and R_4 is 800 ohms. The following method is used to determine the potential difference between points C and D.

1. We have to determine what the voltage is between A and C. We also have to determine the voltage between A to D.
2. Current through the branch from A to B by way of R_1 and R_3 is determined simply by using Ohm's law for current: $I=\frac{E}{R}$.

   Since the resistance in this branch is a series of two resistors, you have to add them to obtain their total resistance.
3. So:

$$I_{AC}=\frac{E}{R_1+R_3}$$

4. Substitute values into the formula:

$$I_{AC}=\frac{12}{1000+2000}$$
$$I_{AC}=\frac{12}{3000}$$
$$I_{AC}=0.0004 \text{ A or } 4 \text{ mA}$$

5. Now we have to find the current from A to D. Since R_2 and R_4 are in series, we find the current through this branch, which has 12 volts applied across it, in order to obtain the current through R_2.
6. That means:

$$I_{AD}=\frac{E}{R_2+R_4}$$
$$I_{AD}=\frac{12}{400+800}$$
$$I_{AD}=\frac{12}{1200}$$
$$I_{AD}=0.01 \text{ A or } 10 \text{ mA}$$

7. Now we must find the voltage drop across R_1 and R_2. This will tell us if there is any difference in voltage between C and D to cause current to flow through the meter.
8. Using Ohm's law again, we can find the voltage drops across these two resistors:

$$E=I\times R$$
$$E_{AC}=0.004\times 1000=4 \text{ volts}$$
$$E_{AD}=0.01\times 400=4 \text{ volts}$$

9. That means terminals C and D are 4 volts positive with respect to terminal A. There is no potential difference, so there is no current flow through the meter.
10. That also means a simple ratio exists between the resistance values. The ratio is:

$$\frac{R_1}{R_3}=\frac{R_2}{R_4}$$

Using this ratio, the resistance of R_2 can be determined if the values of the other three resistors are known:

$$R_2=\frac{R_1\times R_4}{R_3}$$

Or, using the example again:

$$R_2=\frac{R_1\times R_4}{R_3}=\frac{1000\times 800}{2000}$$
$$R_2=\frac{800{,}000}{2{,}000}$$
$$R_2=400\ \Omega$$

PROBLEMS

Find the values missing in the following:

1.

R_1	R_2	R_3	R_4
10	5	40	

2.

R_1	R_2	R_3	R_4
1,000	500		2000

3.

R_1	R_2	R_3	R_4
	5K	10K	5K

4.

R_1	R_2	R_3	R_4
400		2,000	500

NOTE: K = Kilo or 1,000.

Appendix

ANSWERS TO PROBLEMS

The answers provided here are for your use in checking your calculators and calculations. Keep in mind, the problems were solved on a Hewlett-Packard HP-35 calculator, and the answers may be slightly different when a TI-30 II is used. The last two digits are not always the same due to the rounding off done by some calculators. If you use a Casio scientific calculator, the last two numbers may vary from these.

Answers to problems, Chapter 2

Page 20

1. 0.010″
2. 10.0 mils
3. 63.2
4. 0.0000078 sq. in.
5. 4.09 Ω
6. 53,500 Ω
7. 274.56 Ω
8. 124 lb.
9. 320.5 lb.
10. 1.4593 A.
11. 18 turns
12. 5476 turns

Pages 27, 29

1. 3.141 Ω
2. 641.705 Ω
3. 0.136 Ω
4. 0.112 Ω
5. 15.72140699 Ω
6. 6.494816372 Ω
7. 50.78125 Ω
8. 3.973299428 Ω
9. 12.93162741 Ω
10. 17.40338862 Ω

Page 33

1. 2200±10%; 1980–2420 Ω
2. 43,000±5%; 40,850–45,150 Ω
3. (+) 2200 Ω
4. (−) 2,280,000
5. 1.5 Ω±20% (No 4th band given)
6. 0.37 Ω±10%

Pages 36, 37

1. 25 V
2. 20 V
3. 100 V
4. 300 V
5. 94 V
6. 1 A
7. 0.1 A
8. 0.01 A
9. 0.1 A
10. 0.01 A
11. 10 Ω
12. 24 Ω
13. 1000 Ω
14. 1000 Ω
15. 0.001 A

Answers to Problems, Chapter 3

Page 40

1. 100 W
2. 500 W
3. 0.0212765957 W
4. 0.00002727272727 W
5. 500 W
6. 40 W
7. 80 W
8. 250 W
9. 0.5 W
10. 47 W
11. 100 W
12. 1,800 W
13. 3,600 W
14. 375 W
15. 2,000 W

Pages 45, 46

1. 20 Ω
2. 30 Ω
3. 70 Ω
4. 20,115 Ω
5. 2 A
6. 94 V
7. 0.016 A
8. 2 A
9. 330 V
10. 40 Ω
11. 0.5 A
12. 300 Ω
13. Voltage increases to 13.3333333 V across each bulb.
14. No current flow in any part of the circuit.
15. Infinite (∞)
16. a. 100 Ω
    b. 200
    c. 330
    d. 490
    e. 680

Pages 50, 51

1. 25 Ω
2. 27.27272727 Ω
3. 90.09009009 Ω
4. 469.5304695 Ω
5. 333.333333 Ω
6. 3 A
7. 8.4 A
8. 4
9. 1 Ω
10. 8
11. $4.833655706 \times 10^{-5}$ A or 48.33655706 μA
12. 40 V
13. 1 A, 0.5 A, 0.25 A
14. 20 Ω
15. 10

Pages 56, 57

1. 20 Ω
2. 30 Ω
3. 42 Ω
4. 45 Ω
5. 10 Ω

Pages 63, 64

1. A = 15 Ω B = 13.67741936 Ω C = 25 Ω
2. $I_{R_1} = 4$ A $E_{R_1} = 40$ V
   $I_{R_2} = 4$ A $E_{R_2} = 24$ V

$I_{R_3}=2.4$ A	$E_{R_3}=96$ V
$I_{R_4}=1.6$ A	$E_{R_4}=96$ V
$I_{R_5}=4$ A	$E_{R_5}=80$ V

Page 71

1. $I_1=0.5017441864$ A
   $I_2=0.3529069768$ A
   $E_{R_1}=5.01744186$ V
   $E_{R_2}=2.117441861$ V
   $E_{R_3}=5.98255814$ V
2. $I_1=0.0284768213$ A
   $I_2=0.0465894039$ A
   $E_{R_1}=113.9072854$ V
   $E_{R_2}=$ 55.90728476 V
   $E_{R_3}=105.0927153$ V

Answers to problems, Chapter 4

Page 76

	E_{mm}	*SHUNT*
1.	1 V	1.0101010101 Ω
2.	1 V	1.001001001 Ω
3.	0.05 V	0.0050005 Ω
4.	0.5 V	0.500050005 Ω

Page 78

1. 99,950 Ω
2. 499,500 Ω
3. 99,000 Ω
4. 1,999,500 Ω

Page 82

1. 10.5 KΩ
2. 4.5 KΩ
3. 13.5 KΩ
4. 31.5 KΩ
5. 9.0 KΩ

Page 86

	I_{mm}	I_{R_x}
1.	0.333 mA	0.666 mA
2.	0.500 mA	0.500 mA
3.	0.666 mA	0.334 mA
4.	0.750 mA	0.250 mA
5.	0.800 mA	0.200 mA

Answers to problems, Chapter 5

Page 96

1. $0.001 \times 60 \times 360 = 21.6°$
2. $0.002 \times 60 \times 360 = 43.2°$
3. $0.003 \times 60 \times 360 = 64.8°$
4. $0.004 \times 60 \times 360 = 86.4°$

Page 104

1. a. 0.666666666 V
   b. 1.333333333 V
2. a. 1
   b. 1

1. c. 2.0 V
   d. 10 V
   e. 100 V
   f. 1000 V
   g. 10,000 V
   h. 1 V
   i. 40 V
   j. 128 V

2. c. 1
   d. 0.1
   e. −20 V
   f. −40 V
   g. −200 V
   h. −8000 V
   i. −5 V
   j. −1000 V

NOTE: All cemf voltage answers should be negative

Pages 110, 111

1. a. 18 H
   b. 11 H
   c. 17 H
   d. 5.03 H
   e. 0.675 H
2. a. 6 H
   b. 10 H
   c. 16 H
   d. 14 H
   e. 2.8 H
3. a. 0
   b. 6 H
   c. 1 H
   d. 5 H
   e. 0
4. a. 0.25 H
   b. 0.5 H
   c. 0.959999999 H
   d. 0.133333333 H
   e. 0.480000000 H
5. a. 2.4H
   b. 1.428571429 H
   c. 2.66666666 H
   d. 2.22222222 H
   e. 1.05 H
6. a. 0.8571428569 H
   b. 1.5 H
   c. 0.375 H
   d. 0.750 H
   e. 1.5 H

Pages 116, 117

1. a. 0.001 sec
   b. 10 Ω
   c. 20 H
   d. 0.005 sec
   e. 0.005 sec
   f. 0.00012 sec
   g. 125 H
   h. 63.82978723 μsec
   i. 30 H
   j. 15 μsec
2. 0.02 sec
3. 0.333333333 A
4. 0.1 A
5. 0.02127659574 μsec
6. 1 μsec
7. 10 H
8. 3.33333333 Ω
9. 0.1 sec
10. 0.0001 sec

Pages 121, 122

1. a. 3141.592654 Ω
   b. 6283.18531 Ω
   c. 9424.77796 Ω
   d. 12,566.37062 Ω
   e. 15,707.96327 Ω
   f. 18,849.55592 Ω
   g. 21,991.14858 Ω
   h. 25,132.74123 Ω
   i. 28,274.33389 Ω
   j. 31,415.92654 Ω
2. a. 1884.955593 Ω
   b. 785.3981635 Ω
   c. 3769.911185 Ω
   d. 1570.796327 Ω
   e. 400 Hz
   f. 0.2 H
   g. 0.1 H
   h. 550 kHz or 550,000 Hz
   i. 1570.796327 Ω
   j. 3141.592654 Ω

Page 128

1. a. 141.4 V
   b. 282.8 V
   c. 424.2 V
   d. 565.6 V

2. a. 70.71 V
   b. 141.4 V
   c. 212.13 V
   d. 282.84 V

3. a. 9.0 A
   b. 4.5 A
   c. 0.0009 A

Answers to problems, Chapter 6

Pages 134, 135

1. a. 0.01 F
   b. 0.001 F
   c. 0.001 F
   d. 0.0001 F
   e. 0.0001 F
   f. 0.00001 F
   g. 0.00001 F
   h. 0.000001 F
   i. 0.000001 F
   j. 0.0000001 F

2. a. 0.001
   b. 0.01
   c. 0.1
   d. 0.0001
   e. 0.001
   f. 0.003
   g. 0.006
   h. 0.015
   i. 0.15
   j. 0.05
   k. 0.5
   l. 0.1

3. a. 1000 V
   b. 100 V
   c. 10 V
   d. 10,000 V
   e. 100 V
   f. 100,000 V
   g. 10,000 V
   h. 1000 V
   i. 200,000 V
   j. 20,000 V

Pages 139, 140

1. a. 8.84×10^{-5} μF or 88.4 pF
   b. 8.84×10^{-5} μF or 88.4 pF
   c. 8.84×10^{-3} μF or 8840 pF
   d. 8.84×10^{-6} μF or 8.84 pF
   e. 1.48512×10^{-5} μF or 14.8512 pF
   f. 5.8933333×10^{-6} μF or 5.8933333 pF
   g. 3.536×10^{-5} μF or 35.36 pF
   h. 0.01326 μF
   i. 0.0884 μF
   j. 4.42×10^{-4} μF or 442 pF

2. a. 0.1272669424 sq. in.
   b. 1.272669424 sq. in.
   c. 12.72669424 sq. in.
   d. 127.2669424 sq. in.
   e. 1272.669424 sq. in.
   f. 598.1546293 sq. in.
   g. 419.98091 sq. in.
   h. 98.98539965 sq. in.
   i. 20.9990455 sq. in.
   j. 8248.783304 sq. in.

Pages 143, 144

1. a. 0.0025 μF
   b. 0.0048 μF
   c. 0.005 μF
   d. 0.05 μF
   e. 0.0083333 μF
   f. 0.90009 μF
   g. 2.5k pF
   h. 25k pF
   i. 4997.501 μF or 4.997501×10^{-9} F or 4.997501 nF
   j. 1.66666666 μF

2. a. 0.22 μF
   b. 2.02 μF
   c. 400 k pF
   d. 110.101 μF
   e. 0.4 μF or 400 k pF
   f. 0.46 μF
   g. 40.04700047 μF
   h. 100 μF
   i. 275 μF
   j. 43.01 μF

Pages 147, 148

1.	Capacitance	WVDC
a.	2.857142857 μF	111.5
b.	0.826446281 μF	1115.0
c.	33.3333333 μF	3000
d.	25 μF	20
e.	100 μF	101.5

2.	Capacitance	WVDC
a.	35 μF	1.5
b.	106 μF	15
c.	300 μF	1000
d.	100 μF	10
e.	400 μF	1.5

Pages 152, 153

1. a. 0.2 sec
   b. 0.002 sec
   c. 20 μsec or 2×10^{-5} sec
   d. 0.47 sec
   e. 4.7 sec
2. 1 sec
3. 1×10^{-3} A or 1 mA
4. 0.6321 mA
5. 0.8646 mA
6. 0.98168 mA
7. 0.99326 mA
8. 0.05 sec
9. 0.01 sec
10. 0.6321 mA

Pages 154, 155

1. 265.258234 Ω
2. 159,154.9431 Ω
3. 1×10^{-9} F or 0.001 μF
4. $3.29999999 \times 10^{-10}$ F or 329 pF
5. 1×10^{-4} F or 100 μF
6. a. 1591.549431 Ω
   b. 159.1549431 Ω
   c. 15.91549431 Ω
   d. 1.591549431 Ω
   e. 0.1591549431 Ω
   f. 7.442363483 Ω
   g. 0.0615686433 Ω
   h. 30.14298164 Ω
   i. 48.22877062 Ω
   j. 265,258.2384 Ω

Pages 160, 161

1. 1 mA or 0.001 A
2. 5×10^{-4} A or 0.5 mA
3. $3.33333333 \times 10^{-4}$ or 0.33333333 mA
4. 3×10^{-6} A or 3 μA
5. 5×10^{-8} A or 0.05 μA
6. 0.5 A
7. 0.125 A
8. 0.05 A
9. 0.125 A
10. 0.1 A
11. 100,000 V
12. 10,000 V
13. 20,000 V
14. 100,000 V
15. 500,000 V

Answers to problems, Chapter 8

Pages 180, 181

	E_A	I_T	E_L	E_R	Z	R	X_L	$\angle\theta$
a.	141.4213562 V	1 A	100 V	100 V	141.4213562 Ω	100 Ω	100 Ω	45°
b.	1019.803904 V	2 A	200 V	1000 V	509.9019521 Ω	500 Ω	100 Ω	11.30993283°
c.	3354.10965 V	3 A	3000 V	1500 V	1581.138827 Ω	1500 Ω	500 Ω	18.43494851°
d.	4079.21561 V	4 A	4000 V	800 V	1019.803903 Ω	1000 Ω	200 Ω	11.30993254°
e.	111.8033989 V	5 A	50 V	100 V	22.36067977 Ω	10 Ω	20 Ω	63.43494881°
f.	134.1640787 V	6 A	120 V	60 V	22.36067977 Ω	20 Ω	10 Ω	26.56505118°
g.	376.9615366 V	7 A	350 V	140 V	53.85164807 Ω	50 Ω	20 Ω	21.80140948°
h.	288.4441023 V	8A	240 V	160 V	36.05551275 Ω	30 Ω	20 Ω	33.69006752°
i.	648.9992294 V	9 A	360 V	540 V	72.11102551 Ω	40 Ω	60 Ω	56.30993247°
j.	335.4101968 V	10 A	150 V	300 V	33.54101968 Ω	15 Ω	30 Ω	63.43494884°

Pages 185, 186

1. 78.46304097°
2. 48.1896851°
3. 2 A
4. 0.2 A
5. 120 V. Same throughout since it is a parallel circuit.

6.	E_A	I_T	Z	I_R	I_L	R	X_L	$\angle\theta$	E_R	E_L
a.	100 V	2.828427125 A	35.35533906 Ω	2 A	2 A	50 Ω	50 Ω	45°	100 V	100 V
b.	200 V	5.830951895 A	34.29971703 Ω	3 A	5 A	66.66666666 Ω	40 Ω	59.03624347°	200 V	200 V
c.	300 V	7.211102551 A	41.60251472 Ω	4 A	6 A	75 Ω	50 Ω	56.30993247°	300 V	300 V
d.	400 V	11.18033989 A	35.77708763 Ω	5 A	10 A	80 Ω	40 Ω	63.43494883°	400 V	400 V
e.	500 V	13.41640786 A	37.26779964 Ω	6 A	12 A	83.33333333 Ω	41.66666667 Ω	63.43494883°	500 V	500 V
f.	600 V	15.65247585 A	38.33259388 Ω	7 A	14 A	85.71428571 Ω	42.85714286 Ω	63.43494883°	600 V	600 V
g.	700 V	17.0	41.17647059 Ω	8 A	15 A	87.5 Ω	46.66666667 Ω	61.92751307°	700 V	700 V
h.	800 V	13.45362405 A	59.46353169 Ω	9 A	10 A	88.88888888 Ω	80 Ω	48.01278751°	800 V	800 V
i.	900 V	15.62049925 A	57.61659598 Ω	10 A	12 A	90 Ω	75 Ω	50.1944286°	900 V	900 V
j.	1000 V	22.82542442 A	43.81079544 Ω	11 A	20 A	90.90909091 Ω	50 Ω	61.18920626°	1000 V	1000 V

Pages 193, 194

1. 250 Ω
2. 249.9996916 Ω
3. 24.99999969 Ω
4. 1596.94231 Ω
5. 2887.275323 Ω
6. 0.4000000047 A
7. 0.40000049 A
8. 4.000000049 A
9. 0.0751436035 A
10. 4.147267019 A
11. 0.4 A
12. 0.4 A
13. 4.0 A
14. 0.06 A
15. 0.055 A
16. 0.62838315307 μA
17. 0.62838315307 μA
18. 0.62838315307 μA
19. 0.0452389342 A
20. 4.146902304 A

21.

	E_A(V)	I_T(A)	I_R(A)	I_C(A)	F (Hz)	C (μF)		R (Ω)	Z (Ω)	X_C (Ω)
a.	100	1.000000197	1	$6.283185307 \times 10^{-4}$	100	0.01	a.	100	99.9999803	159154.9431
b.	200	0.2015729513	0.2	0.0251327412	200	0.1	b.	1000	992.1966152	7957.747151
c.	300	0.3052830685	0.3	0.0565486677	300	0.1	c.	1000	982.694525	5305.16477
d.	50	0.15715919	0.005	0.1570796327	50	10.0	d.	10,000	318.148751	318.3098861
e.	10	$6.632265135 \times 10^{-4}$	0.0001	$6.283185307 \times 10^{-4}$	10k	0.001	e.	100,000	15,717.67254	15,915.49431

Pages 198, 199

1. 47,000.01074 Ω
2. 470,000.1061 Ω
3. 4,700,001.093 Ω
4. 4,700,107.802 Ω
5. $4.680849994 \times 10^{-3}$ A
6. $4.680850007 \times 10^{-4}$ A
7. $4.680849975 \times 10^{-5}$ A
8. $4.680743704 \times 10^{-5}$ A
9. $\angle\theta = 0.0033732976°$
10. $\angle\theta = 0.038494917°$
11. $\angle\theta = 0.039078927°$
12. $\angle\theta = 0.388059561°$
13. 31.83098861 Ω
14. 318.3098861 Ω
15. 3183.098861 Ω
16. 31830.98861 Ω
17. Very little

Pages 203, 204

	Phase Angles	E_A
1.	a. 45°	141.4213562 V
	b. 63.43494882°	111.8033989 V
	c. 63.43494882°	55.90169947 V
	d. 63.43494882°	11.18033989 V
	e. 26.56505118°	1118.033988 V
	f. 11.30993249°	509.9019514 V
	g. 18.43494851°	1581.138827 V
	h. 36.86989764°	2500 V
	i. 78.69006752°	101.9803901 V
	j. 87.13759479°	100.1249213 V

	Phase Angles	E_A
2.	a.	141.4213562 V
	b.	100 V
	c. 84.28940687°	1004.987563 V
	d. 26.56505111°	55.90169942 V
	e. 63.43494883°	11.18033989 V
	f. 56.30993247°	36.05551275 V
	g. 56.30993247°	18.02775637 V
	h. 63.43494884°	335.4101968 V
	i. 63.43494887°	559.0169954 V
	j. 30°	100.00000 V

Answers to problems, Chapter 9

Page 208

1. a. 60.16491417
   b. 19.02265413
   c. 190.2265413
   d. 1902.265413
   e. 42.53594774
   f. 13.45104773
   g. 1591.549431
   h. 15915.49431
2. It increased.
3. It decreased.
4. Increase L causes decrease in f_r.

Page 227

1.	Frequency	X_L	X_C	I_L	I_C	$\angle\theta$	Applied Voltage
a.	60	3969.911185	265.2582384	.031909964	.031909964	89.84764097	120
b.	50	3141.592654	318.3098861	.0779229846	.0779229846	89.79706058	220
c.	400	25132.74123	39.78873577	.004782218846	.004782218846	89.97716658	120
d.	400	25132.74123	39.78873577	.004383704103	.004383704103	89.97716658	110
e.	25	1570.7969327	636.6197723	0.1230959357	0.1230959357	89.38669459	115
f.	25	1570.7969327	636.6197723	0.2461902529	0.2461902529	89.38669459	230

2.	Frequency	X_L	X_C	I_L	E_L	E_C	E_R	Applied Voltage
a.	60	1884.955593	26.52582384	.0636682818	120.0118839	1.688853627	63.6682818	120
b.	60	1884.955593	26.52582384	.1137228612	214.3625433	3.016592583	113.7228612	240
c.	60	1884.955593	26.52582384	.2084919122	392.997996	5.530923151	208.4919122	440
d.	60	1884.955593	26.52582384	.4169838243	785.9959918	11.06083947	416.9838243	880
e.	50	1570.796327	31.83098861	.119869841	188.291106	3.815575544	119.869841	220
f.	50	1570.796327	31.83098861	.1307670983	205.4084793	4.162446048	130.7670993	240
g.	25	785.3981635	63.66197723	.0891952874	70.05381493	5.678348357	89.19528742	110
h.	25	785.3981635	63.66197723	.1783905748	140.1076298	11.35669671	178.3905748	220

Page 232

E_A (V)	Z (Ω)	I_T (A)	∠θ (degrees)	R (Ω)	X_L (Ω)	X_C (Ω)
100	44.72135956	2.236067977	63.43494881	100	50	25
50	44.72135954	1.118033989	63.43494883	100	25	50
25	2000	0.0125	0	2000	2500	2500
10	4.8507125	2.061552813	14.03624346	5	10	20

Page 245

E_A (V)	I_T (A)	X_L (Ω)	X_C (Ω)	R (Ω)	AP (VA)	TP (W)	∠θ (degrees)	Z (Ω)
100	2.828427125	25	50	25	282.8427125	200	45	35.35533906
50	0.4850712503	50	25	100	24.25356252	23.52941179	14.0362434	103.0776406
25	2.236067977	10	5	10	55.90169843	50	26.56505118	11.18033989

Answers to problems, Chapter 10

Page 262

E_S (volts)	I_P (amps)
360	1
720	1

Pages 268, 269

1. 120 V
2. 208
3. 208
4. Advantage is current
5. Delta
6. Wye
7. 120 V
8. Delta
9. Delta
10. Wye

Page 274

1. $R_X = 65{,}000\ \Omega$
   $R_Y = 13{,}000\ \Omega$
   $R_Z = 6500\ \Omega$
2. $R_a = 6{,}666.666666\ \Omega$
   $R_b = 3{,}333.333333\ \Omega$
   $R_c = 1{,}111.111111\ \Omega$

Page 278

1. 20 Ω
2. 4000 Ω
3. 10 KΩ
4. 100 Ω

CALCULATORS

The only calculator needed for work in this text is simple in use. It calls for trig tables, reciprocals, and square roots. More complicated operations are not necessary. In this book you will be able to do all the problems with the Hewlett-Packard 35 or the Texas Instrument TI-30. Other models are more complicated and expensive and do not necessarily add to the rapidity of solution. They can be used if they are already in your possession. However, it is not necessary to purchase anything with a greater capability than the TI-30 or HP-35 to work all the problems in this book.

Differences in Calculators

The Hewlett-Packard varies slightly from the Texas Instrument calculator. The Hewlett-Packard has "polish notation." This means you enter the first number and then follow the formula as written. The Texas Instrument procedure is slightly different. It is suggested that you follow the instruction booklet closely before attempting to perform the calculations.

The calculator you use may have scientific tables, with the ability to do the following:

Square Root	[$\sqrt{x}$]
Reciprocals	[1/x]
Squares	[X^y]
Trig Functions (especially Cosine)	[cos]
Add	[+]
Subtract	[−]
Divide	[÷]
Multiply	[×]
At least one level of memory	

It would be very helpful to have a calculator that computes pi. The problems in this text have been figured with pi taken to nine places past the decimal. There will be slight variations if you choose to use 3.14 or 3.14159 as pi. [π]

Example 1

A circuit with a resistor and capacitor are connected in series. The resistor has a resistance of 1000 ohms while the capacitor has a capacitance of 0.01 μF. The circuit has 100 volts applied at 100 Hz. What is the phase angle generated by adding the capacitor to the circuit?

Given: $R = 1000$ ohms

$C = 0.01$ μF

$F = 100$ Hz

$E = 100$ V

Find: Phase Angle

1. In order to find the phase angle, you must know the *impedance* (Z) since the

$$[\cos]\angle\theta = \frac{R}{Z}$$

2. Impedance

$$Z = \sqrt{R^2 + X_C^2}$$

Since you don't have the X_C it must be found.

3.
$$X_C = \frac{1}{2\pi FC}$$

4. Use your calculator to make the following calculations:

(Calculations here made on a Hewlett-Packard 35)

$X_C =$ [CLR] [π]

[2][×] [·] [0] [0] [0] [0]

[0] [0] [0] [1] [×] [1] [0] [0] [×] [1/x]

The answer is: 159154.9431 ohms.

[CLR]	clears everything	[1/x]	reciprocal
[STO]	storage	[x^y]	square
[RCL]	recalls from storage	[$\sqrt{x}$]	square root

[SIN] [sine] [COS] [cosine] [TAN] [tangent]

5. Apply X_C in the impedance formula or

$$Z = \sqrt{R^2 + X_C^2}$$

(Calculations here made on a Hewlett-Packard 35)

Z= [CLR] [2] [ENTER] [1] [0] [0] [0] [X^y]

STO 2 ENTER 1 5 9 1 5 4 ·

9 4 3 1 X^y RCL + $\sqrt{x}$

Answer: 159158.0844 ohms

6. You now have the Z and R, so the phase angle can be found.

$$\text{Cos}\angle\theta = \frac{R}{Z}$$

(Calculations here made on a Hewlett-Packard 35)

Cos ∠θ = CLR 1 0 0 0 ENTER 1

5 9 1 5 8 · 0 8 4 4 ÷

ARC COS Answer 89.64000473° CLR OFF

ELECTRICAL AND ELECTRONIC FORMULAS

Resistance

Series Resistors

$$R_T = R_1 + R_2 + R_3 + \cdots$$

$$I_T = I_{R_1} = I_{R_2} = I_{R_3} \cdots$$

$$E_A = E_{R_1} + E_{R_2} + E_{R_3} + \cdots$$

Parallel Resistors

$$R_T = \frac{R_1 \times R_2}{R_1 + R_2} \quad \text{(for TWO resistors } \textit{only}\text{)}$$

$$\frac{1}{R_T}=\frac{1}{R_1}+\frac{1}{R_2}+\frac{1}{R_3}+\cdots$$

$$I_T=I_{R_1}+I_{R_2}+I_{R_3}+\cdots$$

$$E_A=E_{R_1}=E_{R_2}=E_{R_3}$$

$$R=\frac{\rho\times L}{A}$$

Ohm's Law

$$E=I\times R \qquad I=\frac{E}{R} \qquad R=\frac{E}{I}$$

Power

$$P=E\times I \qquad P=I^2R \qquad P=\frac{E^2}{R}$$

Inductance

$$L=\frac{0.4\pi N^2\mu A}{1}\times 10^{-8}$$

Parallel

$$L_T=\frac{1}{\frac{1}{L_1}+\frac{1}{L_2}+\frac{1}{L_3}+\cdots} \quad \text{or} \quad \frac{1}{L_T}=\frac{1}{L_1}+\frac{1}{L_2}+\frac{1}{L_3}$$

$$\text{or} \quad \text{(2 only)} \; L_T=\frac{L_1\times L_2}{L_1+L_2}$$

Series

$$L_T=L_1+L_2+L_3+\cdots$$

Mutual Inductance

$$M=\frac{L_A-L_B}{4} \qquad L_A \;\text{Coils aiding}$$

L_B Coils opposing

$L_T = L_1 + L_2 + 2M$ Series Aiding

$L_T = L_1 + L_2 - 2M$ Series opposing

Time Constant

$$T = \frac{L}{R}$$

Inductance

$$T = R \times C$$

Capacitive

Alternating Current

Peak = rms × 1.414

Average = peak × 0.637

rms = peak × 0.7071

Peak-to-peak = rms × 2.828

Inductive Reactances

$$X_L = 2\pi FL$$

In Series

$$X_{L_T} = X_{L_1} + X_{L_2} + \cdots$$

In Parallel

$$X_{L_T} = \frac{X_{L_1} \times X_{L_2}}{X_{L_1} + X_{L_2}}$$

$$\frac{1}{X_{L_T}} = \frac{1}{X_{L_1}} + \frac{1}{X_{L_2}} + \frac{1}{X_{L_3}}$$

Transformers

$$\frac{E_P}{E_S} = \frac{I_S}{I_P}$$

$$\frac{T_P}{T_S}=\frac{E_P}{E_S}$$

I_P = current, primary
I_S = current, secondary
E_P = voltage, primary
E_S = voltage, secondary
T_P = turns, primary
T_S = turns, secondary

Capacitances

Capacitors in Series

(All in the same unit of measurement)

$$\frac{1}{C_T}=\frac{1}{C_1}+\frac{1}{C_2}+\frac{1}{C_3}+\cdots$$

$$C_T=\frac{C_1\times C_2}{C_1+C_2}\text{ (2 only)}$$

Capacitors in Parallel

(All in the same unit of measurement)

$$C_T=C_1+C_2+C_3+\cdots$$

Capacitance

$$C=8.84\times 10^{-8}K\frac{A}{d}\text{(cm)}$$

$$C=22.45\times 10^{-8}K\frac{A}{d}\text{(inch)}$$

$$C=\frac{Q}{E}$$

$$Q=\frac{A}{d}$$

K = dielectric constant
C = capacitance
Q = charge
d = distance between plates
E = voltage
A = area of plate

Capacitive Reactances

$$X_C = \frac{1}{2\pi FC}$$

In Series

$$X_{C_T} = X_{C_1} + X_{C_2} + X_{C_3}$$

In Parallel

$$\frac{1}{X_{C_T}} = \frac{1}{X_{C_1}} + \frac{1}{X_{C_2}} + \frac{1}{X_{C_3}} + \cdots$$

$$X_{C_T} = \frac{X_{C_1} \times X_{C_2}}{X_{C_1} + X_{C_2}} \text{ (For TWO } \textit{only}\text{)}$$

Ohm's Law for Alternating Current (a.c.) Circuits

$$E_L = I_L \times X_L \qquad E_C = I_C \times X_C$$

$$I_L = \frac{E_L}{X_L} \qquad I_C = \frac{E_C}{X_C}$$

$$X_L = \frac{E_L}{I_L} \qquad X_C = \frac{E_C}{I_C}$$

Series RL Circuit

$$Z = \sqrt{R^2 + X_L^2}$$

$$E_A = \sqrt{E_R^2 + E_L^2}$$

$$I_T = I_R = I_L$$

$$E_A = I_T Z$$

$$*PF = \text{Cos}\angle\theta$$

$$\text{Cos}\angle\theta = \frac{R}{Z} = \frac{E_R}{E_A}$$

$$**TP = AP \times PF$$

$$TP = \text{Watts}$$

$$***AP = \text{Volt-Amperes}$$

$$\text{Phase } \angle = \angle\theta$$

*PF = Power factor

**TP = True power

***AP = Apparent power

Series RC Circuit

$$Z = \sqrt{R^2 + X_C^2}$$

$$E_A = \sqrt{E_R^2 + E_C^2}$$

$$I_T = I_R = I_C$$

$$E_A = I_T Z$$

$$PF = \text{Cos}\angle\theta$$

$$\text{Cos } \angle\theta = \frac{R}{Z} = \frac{E_R}{E_A}$$

$$TP = AP \times PF$$

$$TP = \text{Watts}$$

$$AP = \text{Volt-Amperes}$$

$$\text{Phase } \angle = \angle\theta$$

Series RCL Circuits

$$Z = \sqrt{R^2 + (X_L - X_C)^2}$$

$$E_A = \sqrt{E_R^2 + (E_L - E_C)^2}$$

$$I_T = I_R = I_L = I_C$$

$$\text{Cos}\angle\theta = \text{Power factor } (PF)$$

$$\text{Cos}\angle\theta = \frac{R}{Z} = \frac{E_R}{E_A}$$

$$E_A = I_T Z$$

$$\text{TP} = \text{AP} \times \text{Cos}\angle\theta$$

$$\text{AP} = V \times A$$

Series LC Circuits

$$X_L = X_C \text{ Resonance}$$

$$E_L = E_C \text{ Resonance}$$

$$I_T = \text{Maximum } (\infty)$$

$$Z = \text{Minimum } (0)$$

Coil Merit [No Unit of Measurement]

$$Q = \frac{X_L}{R}$$

Resonance

$$f_r = \frac{1}{2\pi\sqrt{LC}} \qquad \text{or} \qquad f_r = \frac{0.159}{\sqrt{LC}}$$

Parallel RL Circuits

$$Z = \frac{E_A}{I_T}$$

$$E_A = E_R = E_L$$

$$I_T = \sqrt{I_R^2 + I_L^2}$$

$$\text{Cos}\angle\theta = \frac{I_R}{I_T}$$

$$\text{PF} = \text{Cos}\angle\theta$$

$$\text{TP} = \text{AP} \times \text{PF}$$

$$\text{Phase}\ \angle = \angle\theta$$

Parallel RC Circuits

$$Z = \frac{E_A}{I_T}$$

$$E_A = E_R = E_C$$

$$I_T = \sqrt{I_R^2 + I_C^2}$$

$$\text{Cos}\ \angle\theta = \frac{I_R}{I_T}$$

$$\text{PF} = \text{Cos}\ \angle\theta$$

$$\text{TP} = \text{AP} \times \text{PF}$$

$$\text{Phase}\ \angle = \angle\theta$$

Parallel RCL Circuits

$$Z = \frac{E_A}{I_T}$$

$$E_A = E_R = E_L = E_C$$

$$\text{Cos}\ \angle\theta = \frac{I_R}{I_T}$$

$$\text{TP} = \text{AP} \times \text{PF (watts)}$$

$$I_T=\sqrt{I_R^2+(I_L-I_C)^2}$$

Phase $\angle = \angle\theta$

AP = V × A (Volts-Amperes)

PF = Cos $\angle\theta$

Conductance

$$G=\frac{1}{R}$$

$$G_C=2\pi\ FC$$

G = Conductance measured in mho. (Symbol is G but the new unit of measurement is siemen [s] instead of mho.)

Counter EMF (cemf)

$$\text{cemf}=\frac{0.4\pi N^2\mu A}{l}\times\frac{\Delta i}{\Delta t}$$

$$L=\frac{0.4\pi N^2\mu A}{l}\times 10^{-8}$$

$$\text{cemf}=-L\frac{\Delta i}{\Delta t}$$

Three-Phase Power

To convert from Δ to Y:

$$R_a=\frac{R_y\times R_z}{R_x+R_y+R_z}$$

$$R_b=\frac{R_x\times R_z}{R_x+R_y+R_z}$$

$$R_c=\frac{R_x\times R_y}{R_x+R_y+R_z}$$

To convert from Y to Δ:

$$R_x=\frac{(R_a\cdot R_b)+(R_b\cdot R_c)+(R_c\cdot R_a)}{R_a}$$

$$R_y=\frac{(R_a\cdot R_b)+(R_b\cdot R_c)+(R_c\cdot R_a)}{R_b}$$

$$R_z=\frac{(R_a\cdot R_b)+(R_b\cdot R_c)+(R_c\cdot R_a)}{R_c}$$

INDEX

The Audel® Mail Order Bookstore

Here's an opportunity to order the valuable books you may have missed before and to build your own personal, comprehensive library of Audel books. You can choose from an extensive selection of technical guides and reference books. They will provide access to the same sources the experts use, put all the answers at your fingertips, and give you the know-how to complete even the most complicated building or repairing job, in the same professional way.

Each volume:

- **Fully illustrated**
- **Packed with up-to-date facts and figures**
- **Completely indexed for easy reference**

APPLIANCES

HOME APPLIANCE SERVICING, 4th Edition

A practical book for electric & gas servicemen, mechanics & dealers. Covers the principles, servicing, and repairing of home appliances. 592 pages; 5½ × 8¼; hardbound. **Price: $15.95**

REFRIGERATION: HOME AND COMMERCIAL

Covers the whole realm of refrigeration equipment from fractional-horsepower water coolers through domestic refrigerators to multiton commercial installations. 656 pages; 5½ × 8¼; hardbound. **Price: $16.95**

AIR CONDITIONING: HOME AND COMMERCIAL

A concise collection of basic information, tables, and charts for those interested in understanding troubleshooting, and repairing home air-conditioners and commercial installations. 464 pages; 5½ × 8¼; hardbound. **Price: $14.95**

OIL BURNERS, 4th Edition

Provides complete information on all types of oil burners and associated equipment. Discusses burners—blowers—ignition transformers—electrodes—nozzles—fuel pumps—filters—controls. Installation and maintenance are stressed. 320 pages; 5½ × 8¼; hardbound. **Price: $12.95**

AUTOMOTIVE

AUTOMOBILE REPAIR GUIDE, 4th Edition

A practical reference for auto mechanics, servicemen, trainees, and owners. Explains theory, construction, and servicing of modern domestic motorcars. 800 pages; 5½ × 8¼; hardbound. **Price: $14.95**

Use the order coupon on the back of this book.

All prices are subject to change without notice.

AUTOMOTIVE AIR CONDITIONING

You can easily perform most all service procedures you've been paying for in the past. This book covers the systems built by the major manufacturers, even after-market installations. Contents: introduction—refrigerant—tools—air conditioning circuit—general service procedures—electrical systems—the cooling systems—system diagnosis—electrical diagnosis—troubleshooting. 232 pages; $5\frac{1}{2} \times 8\frac{1}{4}$; softcover. **Price: $7.95**

DIESEL ENGINE MANUAL, 4th Edition

A practical guide covering the theory, operation and maintenance of modern diesel engines. Explains diesel principles—valves—timing—fuel pumps—pistons and rings—cylinders—lubrication—cooling system—fuel oil and more. 480 pages; $5\frac{1}{2} \times 8\frac{1}{4}$; hardbound. **Price: $12.95**

GAS ENGINE MANUAL, 2nd Edition

A completely practical book covering the construction, operation, and repair of all types of modern gas engines. 400 pages; $5\frac{1}{2} \times 8\frac{1}{4}$; hardbound. **Price: $9.95**

SMALL GASOLINE ENGINES

A new manual providing practical and theoretical information for those who want to maintain and overhaul two- and four-cycle engines such as lawn mowers, edgers, snowblowers, outboard motors, electrical generators, and other equipment using engines up to 10 horsepower. 624 pp; $5\frac{1}{2} \times 8\frac{1}{4}$; hardbound. **Price: $15.95**

TRUCK GUIDE—3 Vols.

Three all-new volumes provide a primary source of practical information on truck operation and maintenance. Covers everything from basic principles (truck classification, construction components, and capabilities) to troubleshooting and repair. 1584 pages; $5\frac{1}{2} \times 8\frac{1}{4}$; hardbound.
Price: $41.85

Volume 1
ENGINES: **$14.95**
Volume 2
ENGINE AUXILIARY SYSTEMS: **$14.95**
Volume 3
TRANSMISSIONS, STEERING AND BRAKES: **$14.95**

BUILDING AND MAINTENANCE

ANSWERS ON BLUEPRINT READING, 3rd Edition

Covers all types of blueprint reading for mechanics and builders. This book reveals the secret language of blueprints, step by step in easy stages. 312 pages; $5\frac{1}{2} \times 8\frac{1}{4}$; hardbound.
Price: $9.95

BUILDING MAINTENANCE, 2nd Edition

Covers all the practical aspects of building maintenance. Painting and decorating; plumbing and pipe fitting; carpentry; heating maintenance; custodial practices and more. (A book for building owners, managers, and maintenance personnel.) 384 pages; $5\frac{1}{2} \times 8\frac{1}{4}$; hardbound.
Price: $9.95

COMPLETE BUILDING CONSTRUCTION

At last—a one volume instruction manual to show you how to construct a frame or brick building from the footings to the ridge. Build your own garage, tool shed, other outbuildings—even your own house or place of business. Building construction tells you how to lay out the building and excavation lines on the lot; how to make concrete forms and pour the footings and foundation; how to make concrete slabs, walks, and driveways; how to lay concrete block, brick and tile; how to build your own fireplace and chimney. It's one of the newest Audel books, clearly written by experts in each field and ready to help you every step of the way. 800 pages; $5\frac{1}{2} \times 8\frac{1}{4}$; hardbound. **Price: $19.95**

Use the order coupon on the back of this book.

All prices are subject to change without notice.

WELDER/FITTERS GUIDE

Provides basic training and instruction for those wishing to become welder/fitters. Step-by-step learning sequences are presented from learning about basic tools and aids used in weldment assembly, through simple work practices, to actual fabrication of weldments. 160 pages; $8^1/_2 \times 11$; softcover. **Price: $7.95**

FLUID POWER

PNEUMATICS AND HYDRAULICS, 4th Edition

Fully discusses installation, operation and maintenance of both HYDRAULIC AND PNEUMATIC (air) devices. 496 pages; $5^1/_2 \times 8^1/_4$; hardbound. **Price: $15.95**

PUMPS, 4th Edition

A detailed book on all types of pumps from the old-fashioned kitchen variety to the most modern types. Covers construction, application, installation, and troubleshooting. 480 pages; $5^1/_2 \times 8^1/_4$; hardbound. **Price: $14.95**

HYDRAULICS FOR OFF-THE-ROAD EQUIPMENT

Everything you need to know from basic hydraulics to troubleshooting hydraulic systems on off-the-road equipment. Heavy-equipment operators, farmers, fork-lift owners and operators, mechanics—all need this practical, fully illustrated manual. 272 pages; $5^1/_2 \times 8^1/_4$; hardbound.
Price: $8.95

HOBBY

COMPLETE COURSE IN STAINED GLASS

Written by an outstanding artist in the field of stained glass, this book is dedicated to all who love the beauty of the art. Ten complete lessons describe the required materials, how to obtain them, and explicit directions for making several stained glass projects. 80 pages; $8^1/_2 \times 11$; softbound.
Price: $6.95

Electric Motors, 4th Edition. 528 pages; $5\frac{1}{2} \times 8\frac{1}{4}$; hardbound. **Price: $12.95**

Guide to the 1984 National Electrical Code. 672 pages; $5\frac{1}{2} \times 8\frac{1}{4}$; hardbound. **Price: $18.95**

House Wiring, 6th Edition. 256 pages; $5\frac{1}{2} \times 8\frac{1}{4}$; hardbound. **Price: $12.95**

Practical Electricity, 4th Edition. 496 pages; $5\frac{1}{2} \times 8\frac{1}{4}$; hardbound. **Price: $13.95**

Questions and Answers for Electricians Examinations, 8th Edition. 288 pages; $5\frac{1}{2} \times 8\frac{1}{4}$; hardbound. **Price: $12.95**

ELECTRICAL COURSE FOR APPRENTICES AND JOURNEYMEN, 2nd Edition
A study course for apprentice or journeymen electricians. Covers electrical theory and its applications. 448 pages; $5\frac{1}{2} \times 8\frac{1}{4}$; hardbound. **Price: $13.95**

FRACTIONAL HORSEPOWER ELECTRIC MOTORS
This new book provides guidance in the selection, installation, operation, maintenance, repair, and replacement of the small-to-moderate size electric motors that power home appliances and over 90 percent of industrial equipment. Provides clear explanations and illustrations of both theory and practice. 352 pages; $5\frac{1}{2} \times 8\frac{1}{4}$; hardbound. **Price: $15.95**

TELEVISION SERVICE MANUAL, 5th Edition
Provides the practical information necessary for accurate diagnosis and repair of both black-and-white and color television receivers. 512 pages; $5\frac{1}{2} \times 8\frac{1}{4}$; hardbound. **Price: $15.95**

ENGINEERS/MECHANICS/MACHINISTS

MACHINISTS LIBRARY, 4th Edition
Covers the modern machine-shop practice. Tells how to set up and operate lathes, screw and milling machines, shapers, drill presses and all other machine tools. A complete reference library. **Price: $35.85**

Volume 1
Basic Machine Shop. 352 pages; $5\frac{1}{2} \times 8\frac{1}{4}$; hardbound. **Price: $12.95**

Volume 2
Machine Shop. 480 pages; $5\frac{1}{2} \times 8\frac{1}{4}$; hardbound. **Price: $12.95**

Volume 3
Toolmakers Handy Book. 400 pages; $5\frac{1}{2} \times 8\frac{1}{4}$; hardbound. **Price: $12.95**

MECHANICAL TRADES POCKET MANUAL, 2nd Edition
Provides practical reference material for mechanical tradesmen. This handbook covers methods, tools equipment, procedures, and much more. 256 pages; 4×6; softcover. **Price: $10.95**

MILLWRIGHTS AND MECHANICS GUIDE, 3rd Edition
Practical information on plant installation, operation, and maintenance for millwrights, mechanics, maintenance men, erectors, riggers, foremen, inspectors, and superintendents. 960 pages; $5\frac{1}{2} \times 8\frac{1}{4}$; hardbound. **Price: $19.95**

POWER PLANT ENGINEERS GUIDE, 3rd Edition
The complete steam or diesel power-plant engineer's library. 816 pages; $5\frac{1}{2} \times 8\frac{1}{4}$; hardbound. **Price: $16.95**

WELDERS GUIDE, 3rd Edition
This new edition is a practical and concise manual on the theory, practical operation and maintenance of all welding machines. Fully covers both electric and oxy-gas welding. 928 pages; $5\frac{1}{2} \times 8\frac{1}{4}$; hardbound. **Price: $19.95**

Use the order coupon on the back of this book.

All prices are subject to change without notice.

Volume 2
Bricklaying, Plastering Rock Masonry, Clay Tile. 384 pages; $5^1/_2 \times 8^1/_4$; hardbound.
Price: $12.95

PAINTING AND DECORATING

This all-inclusive guide to the principles and practice of coating and finishing interior and exterior surfaces is a fundamental sourcebook for the working painter and decorator and an invaluable guide for the serious amateur or building owner. Provides detailed descriptions of materials, pigmenting and mixing procedures, equipment, surface preparation, restoration, repair, and antiquing of all kinds of surfaces. 608 pages; $5^1/_2 \times 8^1/_4$; hardbound. **Price: $18.95**

PLUMBERS AND PIPE FITTERS LIBRARY, 3rd Edition—3 Vols.

A practical, illustrated trade assistant and reference for master plumbers, journeymen and apprentice pipe fitters, gas fitters and helpers, builders, contractors, and engineers. Explains in simple language, illustrations, diagrams, charts, graphs, and pictures the principles of modern plumbing and pipe-fitting practices. **Price: $32.85**

Volume 1
Materials, tools, roughing-in. 320 pages; $5^1/_2 \times 8^1/_4$; hardbound. **Price: $11.95**
Volume 2
Welding, heating, air-conditioning. 384 pages; $5^1/_2 \times 8^1/_4$; hardbound. **Price: $11.95**
Volume 3
Water supply, drainage, calculations. 272 pages; $5^1/_2 \times 8^1/_4$; hardbound. **Price: $11.95**

THE PLUMBERS HANDBOOK, 7th Edition

A pocket manual providing reference material for plumbers and/or pipe fitters. General information sections contain data on cast-iron fittings, copper drainage fittings, plastic pipe, and repair of fixtures. 330 pages; 4×6 softcover. **Price: $9.95**

QUESTIONS AND ANSWERS FOR PLUMBERS EXAMINATIONS, 2nd Edition

Answers plumbers' questions about types of fixtures to use, size of pipe to install, design of systems, size and location of septic tank systems, and procedures used in installing material. 256 pages; $5^1/_2 \times 8^1/_4$; softcover. **Price: $8.95**

TREE CARE MANUAL

The conscientious gardener's guide to healthy, beautiful trees. Covers planting, grafting, fertilizing, pruning, and spraying. Tells how to cope with insects, plant diseases, and environmental damage. 224 pages; $8^1/_2 \times 11$; softcover. **Price: $8.95**

UPHOLSTERING

Upholstering is explained for the average householder and apprentice upholsterer. From repairing and regluing of the bare frame, to the final sewing or tacking, for antiques and most modern pieces, this book covers it all. 400 pages; $5^1/_2 \times 8^1/_4$; hardbound. **Price: $12.95**

WOOD FURNITURE: Finishing, Refinishing, Repair

Presents the fundamentals of furniture repair for both veneer and solid wood. Gives complete instructions on refinishing procedures, which includes stripping the old finish, sanding, selecting the finish and using wood fillers. 352 pages; $5^1/_2 \times 8^1/_4$; hardbound. **Price: $9.95**

ELECTRICITY/ELECTRONICS

ELECTRICAL LIBRARY

If you are a student of electricity or a practicing electrician, here is a very important and helpful library you should consider owning. You can learn the basics of electricity, study electric motors and wiring diagrams, learn how to interpret the NEC, and prepare for the electrician's examination by using these books.

GARDENING, LANDSCAPING, & GROUNDS MAINTENANCE, 3rd Edition
A comprehensive guide for homeowners and for industrial, municipal, and estate grounds-keepers. Gives information on proper care of annual and perennial flowers; various house plants; greenhouse design and construction; insect and rodent controls; and more. 416 pages; $5\frac{1}{2} \times 8\frac{1}{4}$; hardbound. **Price: $15.95**

CARPENTERS & BUILDERS LIBRARY, 5th Edition (4 Vols.)
A practical, illustrated trade assistant on modern construction for carpenters, builders, and all woodworkers. Explains in practical, concise language and illustrations all the principles, advances, and shortcuts based on modern practice. How to calculate various jobs. **Price: $39.95**

Volume 1
Tools, steel square, saw filing, joinery cabinets. 384 pages; $5\frac{1}{2} \times 8\frac{1}{4}$; hardbound.
Price: $10.95

Volume 2
Mathematics, plans, specifications, estimates. 304 pages; $5\frac{1}{2} \times 8\frac{1}{4}$; hardbound.
Price: $10.95

Volume 3
House and roof framing, layout foundations. 304 pages; $5\frac{1}{2} \times 8\frac{1}{4}$; hardbound.
Price: $10.95

Volume 4
Doors, windows, stairs, millwork, painting. 368 pages; $5\frac{1}{2} \times 8\frac{1}{4}$; hardbound.
Price: $10.95

HEATING, VENTILATING, AND AIR CONDITIONING LIBRARY (3 Vols.)
This three-volume set covers all types of furnaces, ductwork, air conditioners, heat pumps, radiant heaters, and water heaters, including swimming-pool heating systems. **Price: $41.95**

Volume 1
Partial Contents: Heating Fundamentals—Insulation Principles—Heating Fuels—Electric Heating System—Furnace Fundamentals—Gas-Fired Furnaces—Oil-Fired Furnaces—Coal-Fired Furnaces—Electric Furnaces. 614 pages; $5\frac{1}{2} \times 8\frac{1}{4}$; hardbound. **Price: $14.95**

Volume 2
Partial Contents: Oil Burners—Gas Burners—Thermostats and Humidistats—Gas and Oil Controls—Pipes, Pipe Fitting, and Piping Details—Valves and Valve Installations. 560 pages; $5\frac{1}{2} \times 8\frac{1}{4}$; hardbound. **Price: $14.95**

Volume 3
Partial Contents: Radiant Heating—Radiators, Convectors, and Unit Heaters—Stoves, Fireplaces, and Chimneys—Water Heaters and Other Appliances—Central Air Conditioning Systems—Humidifiers and Dehumidifiers. 544 pages; $5\frac{1}{2} \times 8\frac{1}{4}$; hardbound. **Price: $14.95**

HOME-MAINTENANCE AND REPAIR: Walls, Ceilings, and Floors
Easy-to-follow instructions for sprucing up and repairing the walls, ceiling, and floors of your home. Covers nail pops, plaster repair, painting, paneling, ceiling and bathroom tile, and sound control. 80 pages; $8\frac{1}{2} \times 11$; softcover. **Price: $6.95**

HOME PLUMBING HANDBOOK, 3rd Edition
A complete guide to home plumbing repair and installation, 200 pages; $8\frac{1}{2} \times 11$; softcover.
Price: $8.95

MASONS AND BUILDERS LIBRARY, 2nd Edition—2 Vols.
A practical, illustrated trade assistant on modern construction for bricklayers, stonemasons, cement workers, plasterers, and tile setters. Explains all the principles, advances, and shortcuts based on modern practice—including how to figure and calculate various jobs. **Price: $24.90**

Volume 1
Concrete Block, Tile, Terrazzo. 368 pages; $5\frac{1}{2} \times 8\frac{1}{4}$; hardbound. **Price: $12.95**

Use the order coupon on the back of this book.
All prices are subject to change without notice.